Christoph Nolte

Investigations of Heterolytic Reactivity

Christoph Nolte

Investigations of Heterolytic Reactivity
from Fluoro-Substituted Benzhydryl Derivatives to Benzhydryl Fluorides

Südwestdeutscher Verlag für Hochschulschriften

Impressum/Imprint (nur für Deutschland/only for Germany)
Bibliografische Information der Deutschen Nationalbibliothek: Die Deutsche Nationalbibliothek verzeichnet diese Publikation in der Deutschen Nationalbibliografie; detaillierte bibliografische Daten sind im Internet über http://dnb.d-nb.de abrufbar.
Alle in diesem Buch genannten Marken und Produktnamen unterliegen warenzeichen-, marken- oder patentrechtlichem Schutz bzw. sind Warenzeichen oder eingetragene Warenzeichen der jeweiligen Inhaber. Die Wiedergabe von Marken, Produktnamen, Gebrauchsnamen, Handelsnamen, Warenbezeichnungen u.s.w. in diesem Werk berechtigt auch ohne besondere Kennzeichnung nicht zu der Annahme, dass solche Namen im Sinne der Warenzeichen- und Markenschutzgesetzgebung als frei zu betrachten wären und daher von jedermann benutzt werden dürften.

Verlag: Südwestdeutscher Verlag für Hochschulschriften GmbH & Co. KG
Heinrich-Böcking-Str. 6-8, 66121 Saarbrücken, Deutschland
Telefon +49 681 37 20 271-1, Telefax +49 681 37 20 271-0
Email: info@svh-verlag.de

Approved by: München, LMU, Diss.,2012

Herstellung in Deutschland:
Schaltungsdienst Lange o.H.G., Berlin
Books on Demand GmbH, Norderstedt
Reha GmbH, Saarbrücken
Amazon Distribution GmbH, Leipzig
ISBN: 978-3-8381-3072-9

Imprint (only for USA, GB)
Bibliographic information published by the Deutsche Nationalbibliothek: The Deutsche Nationalbibliothek lists this publication in the Deutsche Nationalbibliografie; detailed bibliographic data are available in the Internet at http://dnb.d-nb.de.
Any brand names and product names mentioned in this book are subject to trademark, brand or patent protection and are trademarks or registered trademarks of their respective holders. The use of brand names, product names, common names, trade names, product descriptions etc. even without a particular marking in this works is in no way to be construed to mean that such names may be regarded as unrestricted in respect of trademark and brand protection legislation and could thus be used by anyone.

Publisher: Südwestdeutscher Verlag für Hochschulschriften GmbH & Co. KG
Heinrich-Böcking-Str. 6-8, 66121 Saarbrücken, Germany
Phone +49 681 37 20 271-1, Fax +49 681 37 20 271-0
Email: info@svh-verlag.de

Printed in the U.S.A.
Printed in the U.K. by (see last page)
ISBN: 978-3-8381-3072-9

Copyright © 2012 by the author and Südwestdeutscher Verlag für Hochschulschriften GmbH & Co. KG and licensors
All rights reserved. Saarbrücken 2012

Publications

(1) Can One Predict Changes from S_N1 to S_N2 Mechanisms?
Phan, T. B.; Nolte, C.; Kobayashi, S.; Ofial, A. R.; Mayr, H. *J. Am. Chem. Soc.* **2009**, *131*, 11392-11401.

(?) Kinetics of the Solvolyses of Fluoro-Substituted Benzhydryl Derivatives: Reference Electrofuges for the Development of a Comprehensive Nucleofugality Scale
Nolte, C.; Mayr, H. *Eur. J.Org. Chem.* **2010**, 1435-1439.

(3) The First Picoseconds in the Life of Benzhydryl Cations: Ultrafast Generation and Chemical Reactions
Sailer, C. F.; Fingerhut, B. P.; Ammer, J.; Nolte, C.; Pugliesi, I.; Mayr, H.; de Vivie-Riedle, R.; Riedle E. In *Ultrafast Phenomena XVII* Eds.: Chergui, M.; Jonas, D.; Riedle, E.; Schoenlein, R. W.; Taylor, A., Oxford University Press, New York, **2011**, pp. 427-429.

(4) Nucleofugality and Nucleophilicity of Fluoride in Protic Solvents
Nolte, C.; Ammer, J.; Mayr, H. *J. Org. Chem.* **2011**, in preparation.

Conference Contributions

S_N1 and S_N2 Reactions: Continuous Spectrum or Parallel Processes?
Poster presentation in Santiago de Compostela, *ICPOC*, July **2008**.

Table of Contents

A. Summary 1

B. Introduction 14

C. Results and Discussion 20

    1. Kinetics of the Solvolyses of Fluoro-Substituted Benzhydryl Derivatives: Reference Electrofuges for the Development of a Comprehensive Nucleofugality Scale 20

        1.1. Introduction 20

        1.2. Results and Discussion 21

        1.3. Correlation Analysis 25

        1.4. Conclusion 30

        1.5. Experimental Section 30

        1.6. Kinetics 47

        1.7. References 58

    2. Nucleofugality and Nucleophilicity of Fluoride in Protic Solvents 60

        2.1. Introduction 60

        2.2. Results 62

        2.3. Discussion 76

        2.4. Conclusion 80

        2.5. Experimental Section 82

        2.6. References 116

    3. Can One Predict Changes from S_N1 to S_N2 Mechanisms? 120

        3.1. Introduction 120

        3.2. Experimental Section 122

        3.3. Results and Discussion 125

        3.4. Conclusion 143

        3.5. Experimental Section, Practical Part 144

        3.6. References 166

4. Leaving Group Dependence of the S_N1/S_N2 Ratio 171
    4.1. Introduction 171
    4.2. Results and Discussion 171
    4.3. Conclusion and Outlook 182
    4.4. Experimental Section 187
    4.5. References 196
5. Nucleofugality of Bromide in Other Aprotic Solvents 197
    5.1. Introduction 197
    5.2. Results and Discussion 198
    5.3. Conclusion and Outlook 206
    5.4. Experimental Section 207
    5.5. References 221

A. Summary

1. Kinetics of the Solvolyses of Fluoro-Substituted Benzhydryl Derivatives: Reference Electrofuges for the Development of a Comprehensive Nucleofugality Scale

A series of *meta*-fluoro-substituted benzhydryl chlorides, bromides, mesylates, and tosylates (**1,3,4,5,7**)-X were prepared (Scheme 1).

Scheme 1. Heterolytic cleavage of benzhydryl derivatives and substrates.

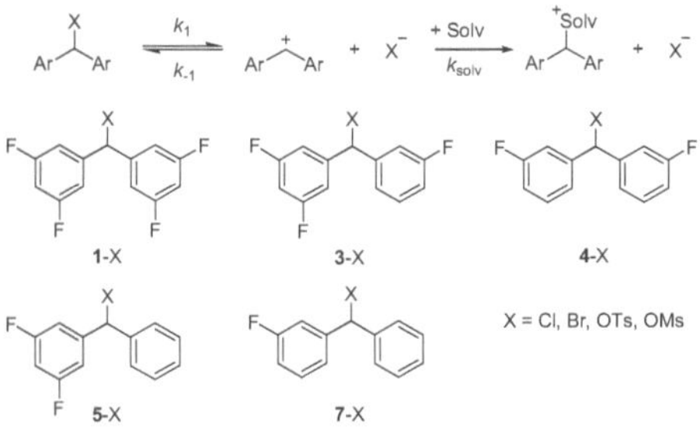

X = Cl, Br, OTs, OMs

The solvolysis reactions of these compounds in various solvents were monitored by conductometry. First-order rate constants were obtained by fitting the time dependent conductances G to the monoexponential function (eq. 1).

$$G = G_\infty(1 - e^{-k_1 t}) + C \quad (1)$$

The obtained first-order rate constants k_1 (25 °C) were found to follow the correlation equation (2) which allowed to determine the electrofugality parameters E_f for these destabilized benzhydrylium cations and the nucleofugality parameters N_f and sensitivity parameters s_f for a series of leaving group-solvent combinations.

$$\lg k_s\,(25\,°C) = s_f(N_f + E_f) \quad (2)$$

A. Summary

For that purpose, the first-order rate constants for the solvolyses of fluorinated benzhydryl derivatives were combined with a large set of solvolysis rate constants of other benzhydryl derivatives and subjected to a least-squares optimization according to equation 3.

$$\sum \Delta^2 = \sum (\lg k_1 - \lg k_{calc})^2 = \sum (\lg k_1 - s_f(N_f + E_f))^2 \quad (3)$$

Initially two fixations were made for this optimization. The electrofugality of the 4,4-dimethoxybenzhydrylium ion (**15$^+$**) was set to zero ($E_f = 0.0$) and the slope for chloride in ethanol was set to one ($s_f = 1.0$). Minimization of the deviation between calculated and experimental rate constants, i.e., $\sum \Delta^2$ as defined by equation 3, yielded the electrofugality parameters for *meta*-fluoro substituted benzhydrylium ions and the nucleofuge-specific parameters N_f/s_f for OTs, OMs, and Br in solvents of high ionizing power (Table 1).

Table 1. Nucleofugality parameters N_f and s_f for leaving groups X in various solvents.

X	OTs	OMs	Br
	N_f / s_f		
TFE	9.73 / 0.94	9.84 / 1.00 [a]	6.19 / 0.95
60AN40W	7.97 / 0.82	7.69 / 0.83	5.23 / 0.99
80E20W	7.44 / 0.80	7.48 / 0.82	4.36 / 0.95
100M	7.33 / 0.82	—	4.23 / 0.99
100E	6.08 / 0.78	5.81 / 0.80	—
80A20W	5.99 / 0.83	5.85 / 0.84	—
90A10W	5.38 / 0.89	—	—

[a] Solvolysis data not included into the total correlation as only two rate constants were available for this leaving-group solvent system.

Thus, eight previously published nucleofugality parameters could be rendered more precisely and seven new nucleofugality parameters were obtained. The nucleofugality parameters can be used to compare leaving-group abilities directly as shown in Figure 1.

A. Summary

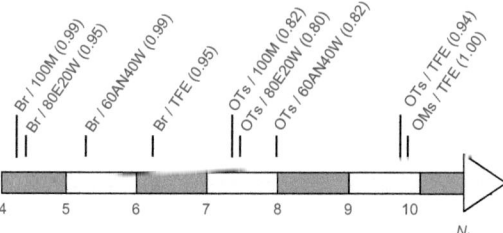

Figure 1. Nucleofugality parameters of bromide, tosylate and mesylate in a series of solvents, slope parameters s_f are given in parentheses.

As shown in Figure 1, tosylate in TFE is a about 6 orders of magnitude better leaving-group than bromide in methanol. Thus, solvolysis reactions of benzhydryl tosylates in TFE will proceed approximately 2×10^5 times faster than analogously substituted benzhydryl bromides in methanol.

A. Summary

2. Nucleofugality and Nucleophilicity of Fluoride in Protic Solvents.

A series of *p*-substituted benzhydryl fluorides (see Scheme 2) were prepared and subjected to solvolysis reactions, which were followed conductometrically.

Scheme 2. Benzhydrylium fluorides employed in this study.

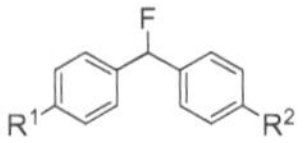

	R^1	R^2
11-F	Me	Me
12-F	OMe	H
13-F	OMe	Me
14-F	OMe	OPh
15-F	OMe	OMe

The observed first-order rate constants k_1 (25 °C) were found to follow the correlation equation (2) and the nucleofuge specific parameters N_f and s_f for fluoride in different aqueous and alcoholic solvents could be determined (Table 2).

Table 2. Nucleofugality parameters N_f and s_f for fluoride in various solvents.

	N_f	s_f
80A20W	−2.72	1.07
80AN20W	−2.28	0.93
100E	−2.21	1.07
60A40W	−2.14	0.81
60AN40W	−1.44	0.84
100M	−1.43	0.99
80E20W	−1.20	0.92

The leaving-group abilities of fluoride are comparable to those of 3,5-dinitrobenzoate. Alkyl fluorides solvolyze approximately 4 to 5 orders of magnitude more slowly than alkyl chlorides, and approximately 5 to 6 orders of magnitude more slowly than alkyl bromides. Figure 2 offers an overview of the solvolytic reactivities of halides and 3,5-dinitrobenzoate.

A. Summary

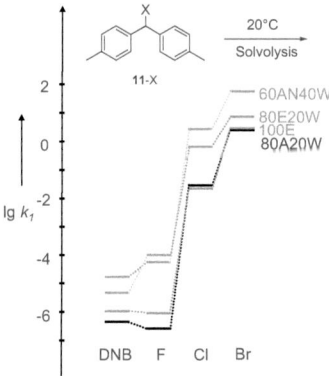

Figure 2. Comparison of the solvolysis rates for the reactions of the dimethyl substituted benzhydryl derivative tol$_2$CH-X (**11-X**) with different leaving groups (DNB = 3,5-dinitrobenzoate). Mixtures of solvents are given as (v/v); solvents: AN = acetonitrile, E = ethanol, W = water.

The nucleophilicity parameters (N, s_N) for fluoride were determined, by generating benzhydrylium ions (diarylcarbenium ions) laser-flash photolytically in various alcoholic and aqueous solvents in the presence of fluoride ions and monitoring the rate of consumption of the benzhydrylium ions by UV-vis spectroscopy. The resulting second-order rate constants k_{-1} (20 °C) and the previously published electrophilicity parameters of benzhydrylium ions were substituted into the correlation equation (4) and the nucleophilicity parameters N and s_N for fluoride in various protic solvents could be derived (Table 3).

$$\lg k_{-1} = s_N(N+E) \qquad (4)$$

A. Summary

Table 3. Nucleophilicity parameters N and s_N for fluoride in various solvents. Mixtures of solvents are given as (v/v); solvents: AN = acetonitrile, M = methanol, W = water.

solvent	N	s_N
90AN10W	12.43	0.58
100M	11.19	0.63
98AN2W	10.88	0.83
80AN20W	10.90	0.61
60AN40W	9.40	0.65
10AN90W	8.05	0.64
100W	7.68	0.65

Fluoride is not only a poorer leaving-group (nucleofuge) than chloride and bromide, but also a poorer nucleophile in protic solvents. The nucleophilicity increases in the series $F^- <$ $Cl^- < Br^-$ in water, aqueous acetonitrile, and methanol as depicted in Figure 3.

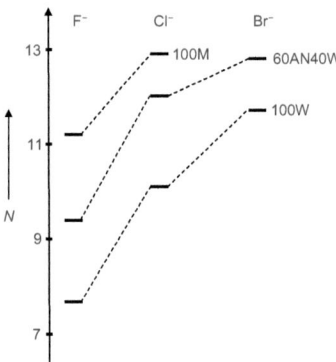

Figure 3. Comparison of the nucleophilic reactivities of fluoride and other halide anions in different solvents. Mixtures of solvents are given as (v/v); solvents: AN = acetonitrile, E = ethanol, M = methanol, W = water.

A. Summary

3. Can One Predict Changes from S_N1 to S_N2 Mechanisms?

Rates and products of the reactions of differently substituted benzhydryl bromides (($\textbf{2,6,8,9,10}$)-Br) with various amines in DMSO (Scheme 3) were studied.

Scheme 3. Reactions of benzhydryl bromides with amines in DMSO.

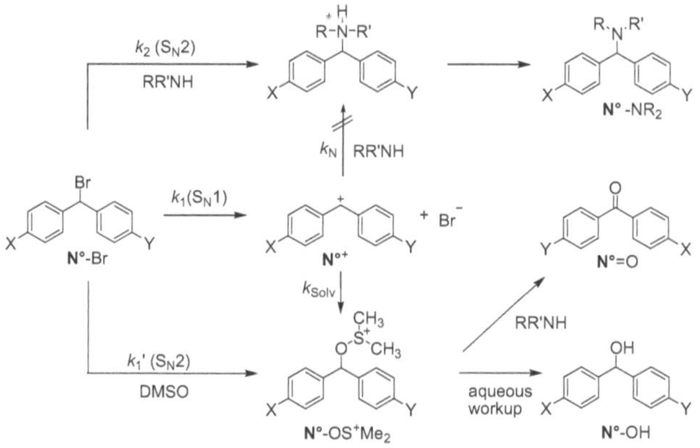

Plots of k_{obs} (first-order rate constants obtained with a large excess of amines) vs. [amine] were linear but did not go through the origin (Figure 4).

A. Summary

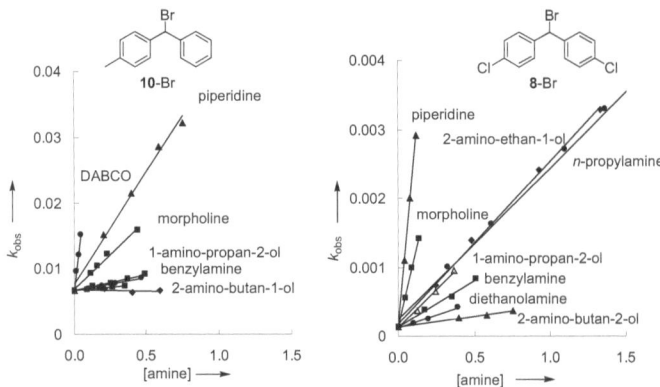

Figure 4. Plots of k_{obs} (s^{-1}) of the reactions of different benzhydryl bromides with amines in DMSO vs. the concentrations of the amines. (Note the different calibration of the y-axes).

The observed rate constants were composed from an amine-independent term k_1 and an amine-dependent term k_2[amine] (eq. 5), indicating parallel S_N1 and S_N2 mechanisms. While the first-order rate constants of the S_N1 reactions correlate linearly with Hammett's substituent constants $\Sigma\sigma^+$, the second-order rate constants k_2 are only weakly affected by the substituents at the aromatic ring and do not correlate significantly with any of Hammett's substituent constants.

$$k_{obs} = k_1 + k_2[\text{amine}] \qquad (5)$$

From the product ratios reported in the last line Table 4, we concluded that the benzhydryl amines (N°-NR$_2$) are exclusively formed via the S_N2 process, because the ratio ([N°-NR$_2$]/([N°=O]+[N°-OH])) is almost identical to the ratio k_2[amine]/k_1. Though the reactions of the benzhydrylium ions (**2$^+$, 6$^+$, 8$^+$, 9$^+$, 10$^+$**) with piperidine are diffusion-controlled they cannot compete with the fast reactions of **N°$^+$** with DMSO, even at a piperidine concentration of 0.2 M.

A. Summary

Table 4. Comparison of the ratios k_2[amine]/k_1 with the product ratio [N°-NR$_2$]/([N°=O]+[N°-OH]) in 0.2 M solution of piperidine in DMSO at 20 °C.

	10-Br	9-Br	8-Br
k_1/s^{-1}	6.71 × 10^{-3}	5.46 × 10^{-4}	1.36 × 10^{-4}
k_2/M^{-1}s^{-1}	3.57 × 10^{-2}	1.69 × 10^{-2}	2.33 × 10^{-2}
k_2 [piperidine]/s^{-1}	7.14 × 10^{-3}	3.38 × 10^{-3}	4.66 × 10^{-3}
k_2 [piperidine]/k_1	1.1	6.2	34
[Ar$_2$CH-NR$_2$]/([Ar$_2$C=O]+[Ar$_2$CH-OH])	1.2	8.7	37

The formation of the benzophenones (N°=O) and benzhydrols (N°-OH) is explained through the intermediacy of the oxysulfonium ions (N°-OS$^+$Me$_2$) which are generated through an S$_N$1 process. Because the ratio [N°=O]/[N°-OH] increases with the reaction time, we conclude that most of the benzhydrol (N°-OH) is generated by hydrolysis of the intermediate oxysulfonium ions (N°-OS$^+$Me$_2$) during aqueous workup.

Both kinetic measurements and product studies show that in the investigated systems the S$_N$1 and S$_N$2 reactions proceed side by side. As the change from S$_N$1 to S$_N$2 mechanisms was observed when the lifetimes of the carbocations in the presence of amines (1 M) were calculated to be approximately 10^{-14} s, Jencks' lifetime criterion was confirmed to be a suitable instrument to predict the change from S$_N$1 to S$_N$2 mechanism.

A. Summary

4. Leaving Group Dependence of the S_N1/S_N2 Ratio

The kinetics of the reactions of the *meta*-fluoro substituted benzhydryl bromides and tosylates depicted in Scheme 4 in DMSO with amines were studied in DMSO in order to investigate the influence of the leaving group on the change from S_N1 to S_N2 mechanism.

Scheme 4. Benzhydryl derivatives employed in this study; E_f parameters are given in parentheses

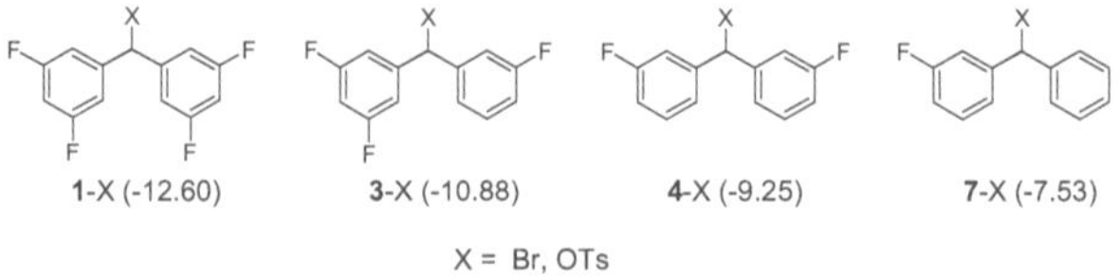

1-X (-12.60)   3-X (-10.88)   4-X (-9.25)   7-X (-7.53)

X = Br, OTs

As in chapter 3, the observed rate constants were composed of an amine-dependent and an amine-independent term (eq. 5) and plots of k_{obs} versus the concentrations of the amines were linear (Figure 4) but did not go through the origin. As depicted for 7-Br and 7-OTs in Figure 5, the benzhydryl tosylates are more reactive in the S_N1 process (larger intercept in Figure 5b) while the benzhydryl bromides react faster by the S_N2 mechanism (larger slopes in Figure 5a).

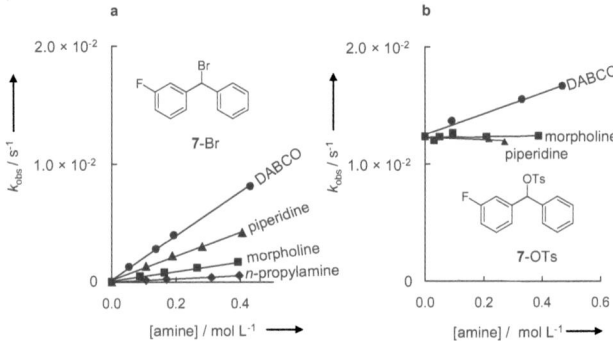

Figure 5. Plots of k_{obs} (s^{-1}) of the reactions of 3-fluorobenzhydryl bromide (7-Br; plots 4a) and tosylate (7-OTs; plots 4b) with amines in DMSO at 20 °C vs. [amine].

A. Summary

The linear plot of lg k_1 for the solvolysis reactions of benzhydryl tosylates with $\Sigma\sigma$ (Figure 6a) suggests that all benzhydryl tosylates investigated solvolyze in DMSO without nucleophilic solvent participation. The corresponding plot for the benzhydryl bromides (Figure 6b) shows a linear correlation for benzhydryl bromides (**4-10**)-Br in line with a nonassisted monomolecular ionization in DMSO. The positive deviation of lg k_1 of (**1-3**)-Br from the correlation line is indicative of a nucleophilic assisted ionization process. The validity of these arguments is questioned by the low sensitivity parameter for tosylate in DMSO (s_f = 0.64) which might be considered as indication of a transition state in which the carbocationic character is not fully developed.

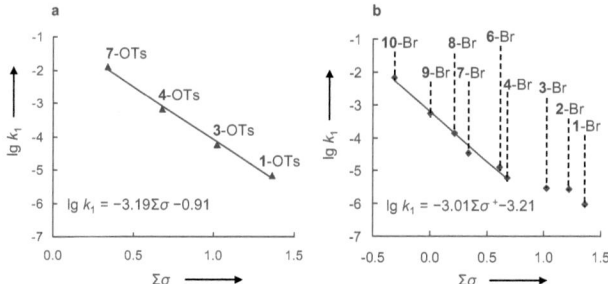

Figure 6. Plot of lg k_1 of the solvolysis reactions of the benzhydryl tosylates (5a) and bromides (5b) in DMSO vs. Hammett's substituent constants σ^+ (σ_m for **1,3,4,7**-Br); benzhydryl bromides with $\Sigma\sigma^+$ > 0.68 were not used for the correlation.

5. Nucleofugality of Bromide in Other Aprotic Solvents

The ionization reactions (k_1) of benzhydryl bromides in aprotic media have been measured by trapping the intermediately formed carbocations by amines (Scheme 5).

Scheme 5. Reversible ionisation of an alkyl halide, followed by fast trapping of the carbocation by a nucleophile and S_N2 reaction with the nucleophile.

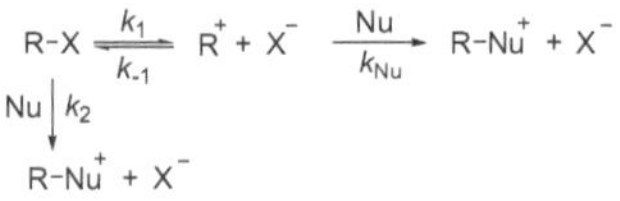

A. Summary

$$R-X \underset{k_{-1}}{\overset{k_1}{\rightleftharpoons}} R^+ + X^- \xrightarrow{\frac{Nu}{k_{Nu}}} R-Nu^+ + X^-$$

$$Nu \downarrow k_2$$

$$R-Nu^+ + X^-$$

Unlike the previously investigated benzhydryl chlorides (Streidl, N.; Mayr, H. *Eur. J. Org. Chem.* **2009**, 2498-2506) the benzhydryl bromides (Scheme 6) used for this investigation do not react exclusively via an S_N1 reaction (k_1), but via parallel S_N1 and S_N2 process.

Scheme 6. Benzhydryl bromides employed in this study.

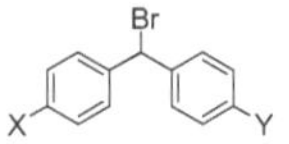

N°-Br

	X	Y
9-Br	H	H
10-Br	Me	H
11-Br	Me	Me
12-Br	OMe	H
13-Br	OMe	Me
14-Br	OMe	OPh
15-Br	OMe	OMe

The observed ionization rate constants k_1, for (**12-15**)-Br, were found to follow the linear free-energy relationship of equation 2, which allowed determining the nucleofugality of bromide in aprotic solvents which are listed in Table 5.

Table 5. Nucleofugality parameters determined by the "amine method" using piperidine and *N*-methylpyrrolidine.

	acetonitrile	DMF	acetone
N_f / s_f	2.09 / 1.45	1.81 / 1.27	1.08 / 1.39

Figure 7 illustrates that the ionization of 4,4'-dimethoxybenzhydryl bromide (**15-Br**) in different aprotic solvents is 400 to 800 times faster than the ionization of the corresponding benzhydryl chloride in the same solvents.

A. Summary

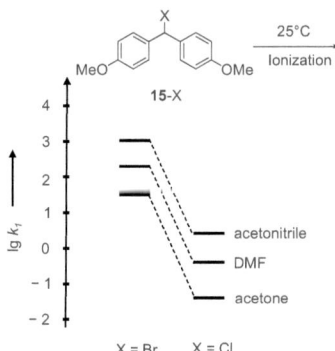

Figure 7. Comparison of calculated lg k_1 for the ionizations of the dimethoxy substituted benzhydryl chloride (**15-Cl**) and bromide (**15-Br**) in three different aprotic solvents.

B. Introduction

Nucleophilic displacement reactions at saturated carbon centers are among the first reactions that are taught in undergraduate lectures. Major concepts in organic chemistry can be illustrated with this simple reaction.

The nucleophilic displacement reaction has been differentiated into S_N1 and S_N2 reactions by Ingold.[1] While in an S_N2 reaction bond formation and breaking proceed simultaneously, the S_N1 reaction proceeds via a stepwise mechanism.

In most textbooks the S_N1 reaction is defined as a reaction where the initial ionization (k_1) is rate determining and the subsequent reaction of the carbocation with the solvent is faster ($k_{solv} > k_{-1}[X^-]$). But this simple reaction scheme is only valid for systems where a highly reactive carbocation is formed. Mechanistic changes can occur when the lifetime of the intermediate carbocation is varied.

Scheme 1. Simplified solvolysis scheme (ion-pair collapse neglected).

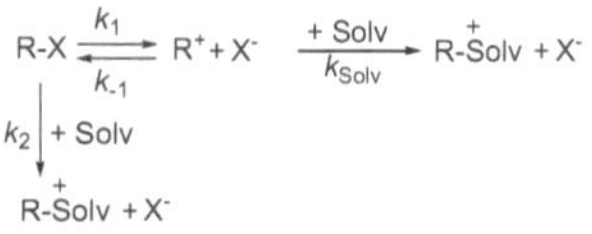

$$R\text{-}X \underset{k_{-1}}{\overset{k_1}{\rightleftharpoons}} R^+ + X^- \xrightarrow[k_{Solv}]{+\,Solv} R\text{-}\overset{+}{S}olv + X^-$$

$$k_2 \downarrow +\,Solv$$

$$R\text{-}\overset{+}{S}olv + X^-$$

The lifetime of the carbocation depends on its electrophilicity and the nucleophilicity of the solvent. Increasing the electrophilicity of the carbocation and nucleophilicity of the solvent will lead to a decreased lifetime of the carbocation. According to Jencks and Richard, a change of mechanism from S_N1 to S_N2 is enforced when the lifetime of the generated carbocation is shorter than a bond vibration ($\approx 10^{-13}$s).[2-5] In such a case, the solvent or a suitable nucleophile may react in an S_N2-type reaction (k_2) with the substrate. When the lifetime of the carbocation is increased, monomolecular reaction will occur ($k_1 > k_2$). If the reaction with the solvent is faster than ion recombination ($k_{solv} > k_{-1}[X^-]$) the ionization step will be rate determining. Further increasing of the lifetime can lead to a solvolysis reaction with common-ion return ($k_{solv} < k_{-1}[X^-]$), S_N2C^+ reactions ($k_{solv} < k_{-1}[X^-]$ and $k_1 > k_{solv}$) and finally to the formation of persistent carbocations ($k_1 > k_{-1}[X^-]; k_1 > k_{solv}$).[6]

For the investigation of these mechanistic changes, a series of substituted benzhydrylium ions (Scheme 2) has been employed previously.[6-10]

B. Introduction

Scheme 2. Benzhydrylium ion used for mechanistic studies.

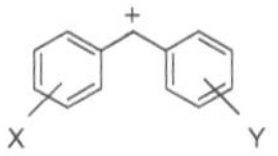

Differently substituted benzhydrylium ions (Table 1) have previously been employed to develop the most comprehensive electrophilicity and nucleophilicity scales presently available.[11-13] Later, benzhydrylium ions have also been employed to construct electrofugality and nucleofugality scales that can be used to predict ionization rate constants of various substrates.[7]

Table 1. Differently substituted benzhydrylium ions employed for the development of comprehensive reactivity scales (not all available benzhydrylium ions from the series of reference electrophiles [11-13] are depicted).

N°⁺	X	Y	E	E_f
1⁺	m-F$_2$	m-F$_2$	8.00 ^{a}	−12.60
2⁺	p-CF$_3$	p-CF$_3$	7.92 ^{a}	-
3⁺	m-F$_2$	m-F	7.58 ^{a}	−10.88
4⁺	m-F	m-F	6.97 ^{a}	−9.26
5⁺	m-F$_2$	H	6.85 ^{a}	−9.00
6⁺	p-CF$_3$	H	6.81 ^{a}	-
7⁺	m-F	H	6.33 ^{a}	−7.53
8⁺	p-Cl	p-Cl	5.59 ^{a}	−6.90
9⁺	H	H	5.60 ^{a}	−6.03
10⁺	p-Me	H	4.50 ^{a}	−4.63
11⁺	p-Me	p-Me	3.63	−3.44
12⁺	p-OMe	H	2.11	−2.09
13⁺	p-OMe	p-Me	1.48	−1.32
14⁺	p-OMe	p-OPh	0.61	−0.86
15⁺	p-OMe	p-OMe	0.00	0.00
16⁺	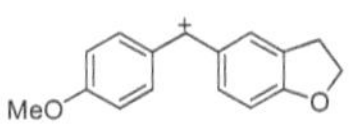		−0.83 ^{a}	0.61

B. Introduction

Table 1. (continued)

N^{o+}		E	E$_f$
17$^+$	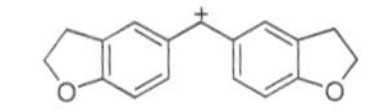 (benzofuran-CH$^+$-dihydrobenzofuran)	−1.36	1.07
18$^+$	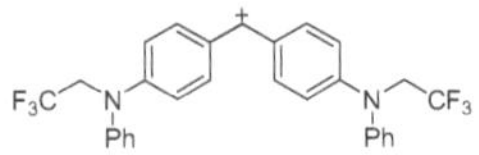 (F$_3$C-N(Ph)-C$_6$H$_4$-CH$^+$-C$_6$H$_4$-N(Ph)-CF$_3$)	−3.14	1.79
19$^+$	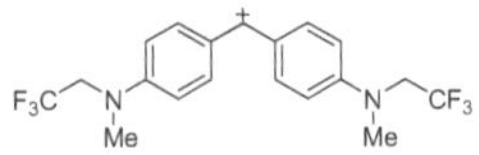 (F$_3$C-N(Me)-C$_6$H$_4$-CH$^+$-C$_6$H$_4$-N(Me)-CF$_3$)	−3.85	3.13
20$^+$	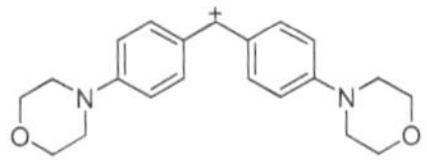 (morpholino-C$_6$H$_4$-CH$^+$-C$_6$H$_4$-morpholino)	−5.53	3.03
21$^+$	NMe$_2$ ... NMe$_2$	−7.02	4.84

^{a} from unpublished work by J. Ammer and C. Nolte

The advantage of benzhydrylium ions is the possibility to vary their reactivities, by different substitutents on the aromatic residue, while the steric environment of the reaction center is kept constant.

One of the most frequently used methods to predict solvolysis rate constants is the Winstein-Grunwald equation (eq. 1).[14,15]

$$\lg (k/k_0) = mY \quad (1)$$

k_0 : rate constant in 80 % aqueous ethanol (80E20W)

m : substrate-specific term (m = 1 for *tert*-butyl chloride)

Y : solvent ionizing power (Y = 0 in 80E20W)

While this correlation provides reliable predictions for the influence of the solvents on the rates of ionizations of C-Cl, C-Br, and C-OTs bonds, it does not compare the nucleofugalities of different leaving groups.

In 2004, equation 2 was introduced as a comprehensive approach to quantify ionization reactions.[21]

B. Introduction

$$\lg k_1(25\,°\mathrm{C}) = s_\mathrm{f}(N_\mathrm{f} + E_\mathrm{f}) \qquad (2)$$

$s_\mathrm{f}, N_\mathrm{f}$: nucleofuge-specific parameters

E_f: electrofuge-specific parameter

In this equation, carbocations are characterized by the electrofugality parameter E_f, while the nucleofuge-specific parameters N_f and s_f refer to combinations of leaving groups and solvents. This equation has been used to quantify 39 reference electrofuges (i.e., carbocations) and over 100 solvent leaving-group combinations.[7] Reference electrofuges that are used in this thesis are depicted in Table 1. Typically, the deviation between calculated and measured solvolysis rate constants was less than 10 %, when reference electrofuges were used. With the reference electrofuges thus defined (see Table 1) it has become possible to characterize the nucleofugality of almost any combination of leaving-group and solvent. Thus, acceptor-substituted benzhydrylium ions are used to characterize the leaving group abilities of good leaving groups (e.g., tosylate), while donor-substituted benzhydrylium ions are used for the characterization of poor leaving groups (e.g., fluoride) as illustrated by the substrates depicted in Figure 1, all of which solvolyze within a couple of minutes and, therefore, can conveniently be measured.

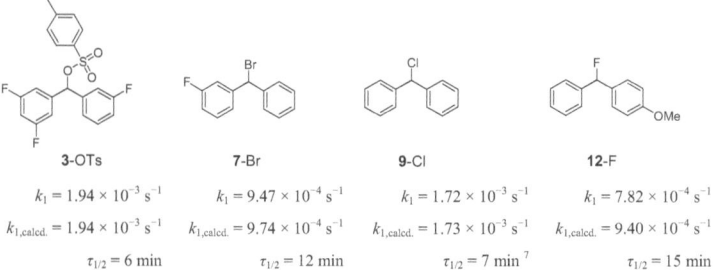

3-OTs	7-Br	9-Cl	12-F
$k_1 = 1.94 \times 10^{-3}\,\mathrm{s}^{-1}$	$k_1 = 9.47 \times 10^{-4}\,\mathrm{s}^{-1}$	$k_1 = 1.72 \times 10^{-3}\,\mathrm{s}^{-1}$	$k_1 = 7.82 \times 10^{-4}\,\mathrm{s}^{-1}$
$k_{1,\mathrm{calcd.}} = 1.94 \times 10^{-3}\,\mathrm{s}^{-1}$	$k_{1,\mathrm{calcd.}} = 9.74 \times 10^{-4}\,\mathrm{s}^{-1}$	$k_{1,\mathrm{calcd.}} = 1.73 \times 10^{-3}\,\mathrm{s}^{-1}$	$k_{1,\mathrm{calcd.}} = 9.40 \times 10^{-4}\,\mathrm{s}^{-1}$
$\tau_{1/2} = 6$ min	$\tau_{1/2} = 12$ min	$\tau_{1/2} = 7$ min [7]	$\tau_{1/2} = 15$ min

Figure 1. A series of benzhydryl derivatives that were employed in solvolytic studies. Observed first-order rate constants, calculated rate constants according to equation 2 and half-lifes of their solvolysis in 80 % aqueous ethanol (80E20W) are displayed below each compound. Rate constant for Ph$_2$CHCl was taken from Ref. [7].

B. Introduction

Initially, *m*-chloro-substituted benzhydryl derivatives were used for the characterization of good leaving-groups (e.g., tosylate). But, they turned out to cause severe skin irritations.[9] Therefore, one goal of this thesis was to replace these substances by *m*-fluoro-substituted benzhydryl derivatives (**1,3,4,5,7**-X) and to use them for characterizing further solvent-leaving group combinations.

In previous investigations, the nucleofugalities of chloride and bromide have been determined in a series of solvents.[7] However, fluoride was so far not incorporated into the nucleofugality scales since the benzhydryl fluorides have not been readily accessible. The determination of the nucleofugality of fluoride in different solvents and the comparison with its nucleophilicity should, therefore, be attempted.

An additional goal of this thesis was to use the newly developed benzhydrylium references for investigating the change from S_N1 to S_N2 mechanism and to examine whether the life-time hypothesis by Richard and Jencks can be used to predict this change.[2-5]

As parts of this thesis have been already published, more detailed introductions are given at the beginning of each chapter.

References:

(1) Ingold, C. K. *Structure and mechanism in organic chemistry*, 2nd ed.; Cornell University Press: Ithaca, NY, 1969.

(2) Jencks, W. P. *Acc. Chem. Res.* **1980**, *13*, 161-169.

(3) Jencks, W. P. *Chem. Soc. Rev.* **1981**, *10*, 345-375.

(4) Richard, J. P.; Jencks, W. P. *J. Am. Chem. Soc.* **1984**, *106*, 1373-1383.

(5) Richard, J. P.; Jencks, W. P. *J. Am. Chem. Soc.* **1984**, *106*, 1383-1396.

(6) Mayr, H.; Ofial, A. R. *Pure Appl. Chem.* **2009**, *81*, 667-683.

(7) Streidl, N.; Denegri, B.; Kronja, O.; Mayr, H. *Acc. Chem. Res.* **2010**, *43*, 1537-1549.

(8) Schaller, H. F.; Mayr, H. *Angew. Chem.* **2008**, *120*, 4022-4025; *Angew. Chem. Int. Ed.* **2008**, *47*, 3958-3961.

(9) Denegri, B.; Streiter, A.; Juric, S.; Ofial, A. R.; Kronja, O.; Mayr, H. *Chem. Eur. J.* **2006**, *12*, 1648-1656.

(10) Denegri, B.; Ofial, A. R.; Juric, S.; Streiter, A.; Kronja, O.; Mayr, H. *Chem. Eur. J.* **2006**, *12*, 1657-1666.

B. Introduction

(11) Mayr, H.; Patz, M. *Angew. Chem. Int. Ed.* **1994**, *33*, 938-957.
(12) Mayr, H.; Bug, T.; Gotta, M. F.; Hering, N.; Irrgang, B.; Janker, B.; Kempf, B.; Loos, R.; Ofial, A. R.; Remennikov, G.; Schimmel, H. *J. Am. Chem. Soc.* **2001**, *123*, 9500-9512.
(13) Mayr, H.; Kempf, B.; Ofial, A. R. *Acc. Chem. Res.* **2003**, *36*, 66-77.
(14) Grunwald, E.; Winstein, S. *J. Am. Chem. Soc.* **1948**, *70*, 846-854.
(15) Fainberg, A. H.; Winstein, S. *J. Am. Chem. Soc.* **1956**, *78*, 2770-2777.
(16) Kwang-Ting, L.; Hun-Chang, S.; Hung-I, C.; Pao-Feng, C.; Chia-Ruei, H. *Tetrahedron Lett.* **1990**, *31*, 3611-3614.
(17) Liu, K.-T.; Chang, L.-W.; Yu, D.-G.; Chen, P.-S.; Fan, J.-T. *J. Phys. Org. Chem.* **1997**, *10*, 879-884.
(18) Bentley, T. W.; Llewellyn, G. In *Prog. Phys. Org. Chem.*; John Wiley & Sons, Inc.: 2007, 121-158.
(19) Kevill, D. N.; D'Souza, M. J. *J. Chem. Res.* **2008**, 61-66.
(20) Winstein, S.; Fainberg, A. H.; Grunwald, E. *J. Am. Chem. Soc.* **1957**, *79*, 4146-4155.
(21) Denegri, B.; Minegishi, S.; Kronja, O.; Mayr, H. *Angew. Chem.* **2004**, *116*, 2353-2356.

1. Kinetics of the Solvolyses of Fluoro-Substituted Benzhydryl Derivatives

C. Results and Discussion

1. Kinetics of the Solvolyses of Fluoro-Substituted Benzhydryl Derivatives: Reference Electrofuges for the Development of a Comprehensive Nucleofugality Scale

1.1. Introduction

The stabilization of benzhydryl cations (diarylcarbenium ions)[1] can be modified widely by variation of substituents in *p*- and *m*-position, while the steric shielding of the carbocation center is kept constant. For that reason, benzhydrylium ions ($N^{\circ+}$) have not only been used to construct the most comprehensive nucleophilicity scale presently available,[2] but also for the development of a nucleofugality scale.[3] In order to compare the leaving group abilities of tosylate and bromide in solvents of high ionizing power, we had studied the solvolyses of mono- to tetra-(*m*-chloro) substituted benzhydrylium derivatives.[3a,b] However, as mentioned in a previous report,[3a] several researchers were suffering from severe skin irritations when working in a laboratory where these compounds were used. For that reason we had to abandon the *m*-chloro substituted compounds as references and replace them by the corresponding fluoro derivatives **1-X**, **3-X**, **4-X** and **7-X** with one to four *m*-fluoro substituents (Scheme 1.1).

The highly electrophilic benzhydrylium ions generated from these precursors have recently been employed to study the fastest bimolecular reactions in the electronic ground state we are aware of, i.e., the reaction of 1^+ with methanol which proceeds with a reaction time of 2.6 ps, corresponding to a time in which light propagates less than 1 mm.[4] We now report on the synthesis and characterization of compounds **1-X**, **3-X**, **4-X**, **5-X**, **7-X** and the kinetics of their solvolysis reactions in order to determine the electrofugality parameters of $1^+, 3^+, 4^+, 5^+$ and 7^+.

1. Kinetics of the Solvolyses of Fluoro-Substituted Benzhydryl Derivatives

Scheme 1.1. Benzhydrylium derivatives employed in this study.

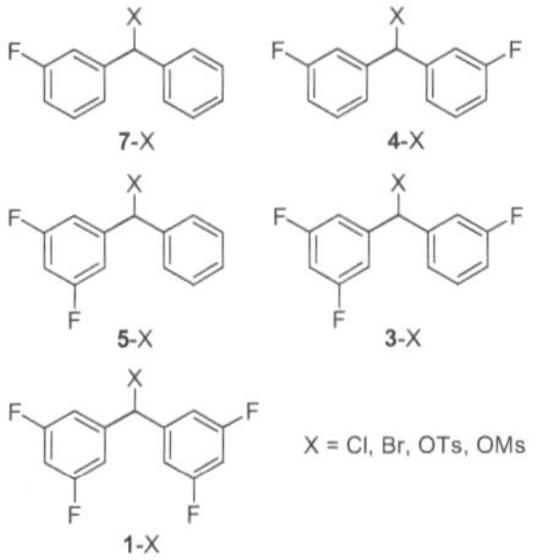

7-X   4-X

5-X   3-X

1-X   X = Cl, Br, OTs, OMs

1.2. Results and Discussion

Synthesis of the precursors: The fluoro substituted benzhydrols (**3,4,5,7**)-OH were synthesized by the reactions of the fluorinated phenylmagnesium bromides with fluorinated benzaldehydes (Table 1.1) as described in detail in the experimental section. For the synthesis of the symmetrical tetrafluoro-substituted benzhydrol **1**-OH, 3,5-difluorophenylmagnesium bromide was combined with 0.5 equivalents of ethyl formate.

Table 1.1. Synthesis of the fluoro-substituted benzhydrols (**1-4**)-OH.

Product	Grignard reagent	Aldehyde reagent
7-OH	PhMgBr	(3-FC_6H_4)CHO
4-OH	(3-FC_6H_4)MgBr	(3-FC_6H_4)CHO
5-OH	(3,5-$F_2C_6H_3$)MgBr	PhCHO
3-OH	(3,5-$F_2C_6H_3$)MgBr	(3-FC_6H_4)CHO
1-OH	(3,5-$F_2C_6H_3$)MgBr	0.5 HC(O)(OEt)

Standard reagents ($SOCl_2$, PBr_3) were used to convert the benzhydrols (**1,3,4,7**)-OH into the benzhydryl chlorides (**1,3,4,7**)-Cl and benzhydryl bromides (**1,3,4,5,7**)-Br. Attempts to convert the benzhydrols (**1,3,4,7**)-OH to the corresponding tosylates with *p*-toluenesulfonyl chloride/pyridine, *p*-toluenesulfonyl anhydride/triethylamine, or *p*-toluenesulfonyl

1. Kinetics of the Solvolyses of Fluoro-Substituted Benzhydryl Derivatives

anhydride/ sodium hydride were unsuccessful.[5] Therefore, the benzhydryl tosylates (**1,3,4,7**)-OTs were synthesized in accordance to a procedure by Cheeseman and Poller[6] by treatment of the corresponding benzhydryl bromides (**1,3,4,7**)-Br with silver tosylate in dichloromethane. The benzhydryl mesylates (**1-4**)-OMs were synthesized in the same manner by treatment of the corresponding benzhydryl bromides with silver mesylate. Only the tetra-fluorinated benzhydryl mesylate **1**-OMs was obtained as a pure solid. The other mesylates (**3,4,7**)-OMs could not be isolated as pure substances and were obtained as yellowish oils that could neither be distilled nor brought to crystallization. Nevertheless it was possible to investigate their solvolysis reactions by using diluted solutions, in analogy to a procedure by Bentley.[7] The benzhydryl sulfonates used in this study are sensitive to moisture, and great care has to be taken to exclude traces of moisture during synthesis and handling of these compounds.

Kinetics: When compounds (**1,3,4,5,7**)-X were dissolved in aqueous or alcoholic media, an increase of conductivity due to the generation of HX was observed. Because calibration experiments showed a linear correlation between conductance (G) and the concentration of HX in the concentration range investigated, the observed mono-exponential increase of conductance (G) (eq. 1.1) indicated the operation of a first-order rate law.

$$G = G_\infty(1 - e^{-k_1 t}) + C \qquad (1.1)$$

The first-order rate constants k_1 (Table 1.2) were obtained by fitting the time-dependent conductances G to the monoexponential function (eq. 1.1). Because all solvolyses studied in this work follow first-order rate laws, common-ion return[8] (k_{-1} in Scheme 1.2) obviously does not occur, and the k_1 values listed in Table 1.2. correspond to the ionization rate constants k_1 defined in Scheme 1.2.

Scheme 1.2. Simplified solvolysis scheme.

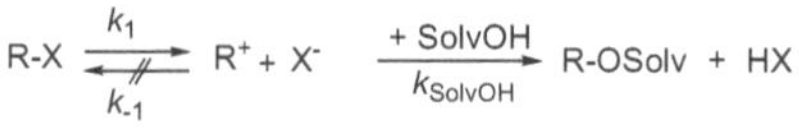

$$\text{R-X} \underset{k_{-1}}{\overset{k_1}{\rightleftharpoons}} \text{R}^+ + \text{X}^- \xrightarrow[k_{\text{SolvOH}}]{+ \text{SolvOH}} \text{R-OSolv} + \text{HX}$$

1. Kinetics of the Solvolyses of Fluoro-Substituted Benzhydryl Derivatives

Table 1.2. Solvolysis rate constants (25 °C) of the benzhydryl derivatives (**1-5**)-X in different solvents.

Solvent[a]	X	electrofuge	k_s/s^{-1}
90A10W	TsO	**7**$^+$	1.16×10^{-2}
		4$^+$	4.24×10^{-4}
80A20W	TsO	**7**$^+$	$5.59 \times 10^{-2\,b}$
		4$^+$	2.42×10^{-3}
		3$^+$	1.10×10^{-4}
80A20W	OMs	**7**$^+$	3.87×10^{-2}
		4$^+$	1.37×10^{-3}
		3$^+$	5.92×10^{-5}
60A40W	Br	**7**$^+$	1.47×10^{-3}
50A50W	Br	**4**$^+$	1.59×10^{-4}
60AN40W	OTs	**7**$^+$	$2.14^{\,b}$
		4$^+$	9.53×10^{-2}
		3$^+$	4.26×10^{-3}
		1$^+$	1.45×10^{-4}
60AN40W	OMs	**4**$^+$	4.82×10^{-2}
		3$^+$	2.52×10^{-3}
		1$^+$	8.04×10^{-5}
60AN40W	Br	**10**$^+$	$4.57^{\,b}$
		9$^+$	$1.44 \times 10^{-1\,b}$
		7$^+$	4.59×10^{-3}
		4$^+$	1.18×10^{-4}
		5$^+$	1.59×10^{-4}
60AN40W	Cl	**7**$^+$	2.67×10^{-4}

1. Kinetics of the Solvolyses of Fluoro-Substituted Benzhydryl Derivatives

Table 1.2. (continued)

Solvent[a]	X	electrofuge	k_s/s^{-1}
100E	OTs	**7**[+]	8.22×10^{-2}
		4[+]	3.35×10^{-3}
		3[+]	1.88×10^{-4}
100E	OMs	**7**[+]	4.34×10^{-2}
		4[+]	1.70×10^{-3}
		3[+]	9.12×10^{-5}
80E20W	OTs	**4**[+]	4.07×10^{-2}
		3[+]	1.94×10^{-3}
		1[+]	8.15×10^{-5}
80E20W	OMs	**4**[+]	3.37×10^{-2}
		3[+]	1.35×10^{-3}
		1[+]	5.27×10^{-5}
80E20W	Br	**7**[+]	9.47×10^{-4}
		4[+]	2.30×10^{-5}
		5[+]	3.98×10^{-5}
100M	OTs	**7**[+]	8.33×10^{-1}
		4[+]	2.07×10^{-2}
		3[+]	1.13×10^{-3}
		1[+]	5.51×10^{-5}
100M	Br	**7**[+]	5.75×10^{-4}
80M20W	Br	**4**[+]	1.90×10^{-4}
100TFE	OTs	**3**[+]	7.99×10^{-2}
		1[+]	1.73×10^{-3}
100TFE	OMs	**3**[+]	9.21×10^{-2}
		1[+]	1.78×10^{-3}

1. Kinetics of the Solvolyses of Fluoro-Substituted Benzhydryl Derivatives

Table 1.2. (continued)

Solvent^a	X	electrofuge	k_s/s^{-1}
100TFE	Br	7$^+$	7.27 × 10^{-2}
		4$^+$	1.49 × 10^{-3}
		5$^+$	2.36 × 10^{-3}
		3$^+$	2.54 × 10^{-5}
100TFE	Cl	7$^+$	2.10 × 10^{-2}
		4$^+$	3.87 × 10^{-4}

^{a} Mixtures of solvents are given as (v/v); solvents: A = acetone, AN = acetonitrile, E = ethanol, M = methanol, TFE = 2,2,2-trifluoroethanol, W = water; ^{b} Stopped flow kinetics.

The Eyring and Arrhenius activation energies were determined for **7-Br** and **3-OTs** in 80E20W as representative systems (Table 1.3). Both compounds exhibit activation parameters which are typical for solvolysis reactions.[3a,c,d, 9]

Table 1.3. Activation parameters for the solvolyses of **1-Br** and **3-OTs** in 80E20W.

	7-Br	3-OTs
$\Delta H^{\ddagger}$/kJ mol^{-1}	84.2 ± 0.7	79.8 ± 0.8
$\Delta S^{\ddagger}$/J mol^{-1} K^{-1}	−20.2 ± 2.4	−28.9 ± 2.7
E_a/kJ mol^{-1}	86.6 ± 2.4	82.2 ± 0.8
lg A	12.2 ± 0.1	11.7 ± 0.1

1.3. Correlation Analysis

In previous work we have demonstrated that equation 1.2 can be used to correlate solvolysis rates of substrates which differ widely in reactivity.[3] In this equation, carbocations are characterized by the electrofugality parameter E_f, while the nucleofuge-specific parameters N_f and s_f refer to combinations of leaving groups and solvents.

1. Kinetics of the Solvolyses of Fluoro-Substituted Benzhydryl Derivatives

$$\lg k_1(25\ °C) = s_f(N_f + E_f) \qquad (1.2)$$

s_f, N_f : nucleofuge-specific parameters
E_f : electrofuge-specific parameter

The solvolysis rate constants k_1 in Table 1.2 and previously reported solvolysis rate constants for benzhydryl derivatives were subjected to a least-squares minimization according to equation 1.3, where E_f for (4-MeOC$_6$H$_4$)$_2$CH$^+$ (**15$^+$**) was set to 0.00 and s_f for the leaving group/solvent combination chloride/ethanol was set to 1.00.

$$\sum \Delta^2 = \sum (\lg k_s - \lg k_{calc})^2 = \sum (\lg k_s - s_f(N_f + E_f))^2 \qquad (1.3)$$

Minimization of the deviation between calculated and experimental rate constants, i.e., $\sum \Delta^2$ as defined by equation 1.3, yielded the electrofugality parameters for the benzhydrylium ions **1$^+$**, **3$^+$**, **4$^+$** and **7$^+$** (Table 1.4.)[3e] and the nucleofuge-specific parameters N_f/s_f for OTs, OMs, and Br in solvents of high ionizing power (Table 1.5.).

Table 1.4. Electrofugality (E_f) parameters of fluoro substituted benzhydryl cations.

benzhydryl cation	substituents	E_f
7$^+$	3-Fluoro	−7.53
4$^+$	3,3'-Difluoro	−9.25
3$^+$	3,3',5-Trifluoro	−10.88
1$^+$	3,3',5,5'-Tetrafluoro	−12.60

1. Kinetics of the Solvolyses of Fluoro-Substituted Benzhydryl Derivatives

Table 1.5. Nucleofugality parameters N_f and s_f for leaving groups X in various solvents.

N_f / s_f			
X	OTs	OMs	Br
TFE	9.73 / 0.94 [a]	9.84 / 1.00 [b]	6.19 / 0.95 [a]
60AN40W	7.97 / 0.82	7.69 / 0.83	5.23 / 0.99
80E20W	7.44 / 0.80 [a]	7.48 / 0.82	4.36 / 0.95 [a]
100M	7.33 / 0.82	—	4.23 / 0.99 [a]
100E	6.08 / 0.78 [a]	5.81 / 0.80	2.93 / 0.93 [c]
80A20W	5.99 / 0.83 [a]	5.85 / 0.84	3.01 / 0.90 [c]
90A10W	5.38 / 0.89 [a]	—	2.29 / 1.01 [c]

[a] These parameters revise previously published values from Ref. [3a]. [b] Solvolysis data not included into the total correlation as only two rate constants were available for this leaving-group solvent system. [c] from Ref. [3e].

Some of the nucleofuge-specific parameters N_f/s_f in Table 1.5 have previously been reported. The small deviation of the new parameters, which are based on a larger set of experimental data, confirms the solidity of the previously reported set of nucleofugality parameters.[3a,b]

Figure 1.1 illustrates the high quality of these correlations for benzhydryl bromides (**A**), tosylates (**B**) and mesylates (**C**).

1. Kinetics of the Solvolyses of Fluoro-Substituted Benzhydryl Derivatives

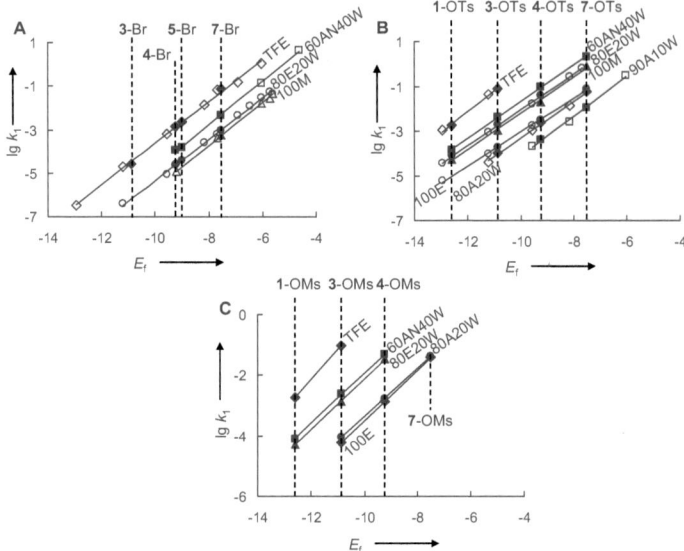

Figure 1.1. Plot of lg k_s for the solvolysis reaction of various substituted benzhydryl derivatives vs. electrofugality E_f for **A** bromides, **B** tosylates, **C** mesylates. Data points with filling were taken from Table 2, data points without filling were taken from previously published data.[3a] Mixtures of solvents are given as (v/v); solvents: A = acetone, AN = acetonitrile, E = ethanol, M = methanol, TFE = 2,2,2-trifluorethanol, W = water.

From the electrofugality parameters in Table 1.4. and the nucleofugality parameters for Cl$^-$ in MeOH (N_f = 2.95, s_f = 0.98)[3a] and EtOH (N_f = 1.87, s_f = 1.00)[3a] one can calculate the methanolysis rate constant for **7**-Cl (3.18 × 10^{-5} s^{-1}) and the ethanolysis rate constant for **7**-Cl (2.14 × 10^{-6} s^{-1}) which agree nicely with the experimental values of 2.97 × 10^{-5} s^{-1} and 1.74 × 10^{-6} s^{-1}, respectively, reported by Nishida.[10b] The closely similar nucleofugalities of TsO and MsO in different solvents is in line with the previously reported similar magnitude of the leaving-group abilities of these two sulfonate groups.[7] The slightly lower nucleofugality of mesylate can be explained by the better delocalization of negative charge by tosylate. Depending on the solvent, the nucleofugality of bromide is 2.5 to 3.8 orders of magnitude smaller.

1. Kinetics of the Solvolyses of Fluoro-Substituted Benzhydryl Derivatives

In line with the similar magnitude of Hammett's σ_m constants for Cl (0.37) and F (0.34),[11] the m-fluoro-substituted benzhydrylium ions 7^+, 4^+, 3^+ and 1^+ have similar electrofugalities as the corresponding chloro-substituted benzhydrylium ions.[3a] Remarkable is the almost constant increment of -1.7 per m-F substituent on the electrofugality E_f of the benzhydrylium ions. In line with this observation, Figure 1.2 shows a linear correlation between E_f and $\sum\sigma_m$ with a slope of -4.91 which corresponds to Hammett ρ values from -4.06 to -5.08 for reaction series with $0.77 < s_f < 0.99$ (Table 1.4). From the three different rate constants for the reaction of 5-Br in TFE, 60AN40W and 80E20W an electrofugality of $E_f = -9.00$ was calculated. The almost identical E_f values of the symmetrical (4^+) and unsymmetrical difluoro-substituted system (5^+) also illustrates the additivity of substituent effects. This behavior contrasts that of donor substituents, where a leveling effect is observed,[10e,12] i.e., the second electron-donor group has generally a smaller cation-stabilizing effect than the first donor group. From the observation that replacement of one H by F has a similar effect in the comparison $3^+ \rightarrow 1^+$ as in the comparison 9^+ (Ph$_2$CH$^+$) $\rightarrow$ 7^+ one may conclude that nonadditivity of substituent effects in benzhydrylium systems is specific for substituents with +M effects.[10] The unsymmetrical difluoro-substituted system 5^+, which exhibits a similar reactivity as the symmetrical difluoro-substituted system 4^+, was not included into the series of reference electrofuges.

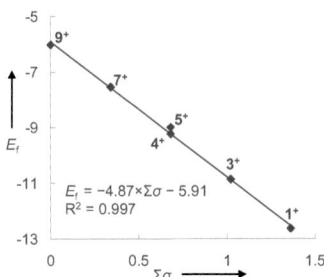

Figure 1.2. Correlation of the electrofugality parameters E_f of benzhydrylium ions (Table 3) with Hammett σ constants (from Ref.[11]).

1. Kinetics of the Solvolyses of Fluoro-Substituted Benzhydryl Derivatives

1.4. Conclusion

The solvolyses of the fluoro-substituted benzhydryl bromides, tosylates and mesylates (**1,3,4,5,7**)-X in various aqueous and alcoholic solutions follow first-order kinetics with rate-determining ionization and no common-ion return.[8] All rate constants follow the correlation lg $k_s = s_f(N_f + E_f)$ (eq. 1.2), which confirms previously reported nucleofugality parameters N_f, s_f for bromide and tosylate in various solvents and allows to determine nucleofugalities for mesylate as well as electrofugalities E_f of the fluoro-substituted benzhydrylium ion 1^+, 3^+, 4^+ and 7^+. The fluorine effects are roughly additive, and the solvolysis rates are retarded by a factor of 18 to 59 per meta-fluorine. Because the E_f values of 1^+, 3^+, 4^+ and 7^+ are similar to those of the corresponding chloro-substituted benzhydrylium ions (Table 1.4), the fluoro-substituted benzhydryl cations 1^+, 3^+, 4^+ and 7^+ are suggested to replace the skin irritant chloro-substituted analogues as references in the high electrophilicity and low electrofugality range for quantifying weak nucleophiles and strong nucleofuges, respectively.

1.5. Experimental Section

Preparation of the benzhydrols (**3,4,5,7**)-OH; general procedure:
In a flame dried, nitrogen-flushed three-necked round-bottom flask, equipped with a reflux condenser and two dropping funnels, magnesium, which was activated with a small amount (tip of a spatula) of iodine at 67 °C, was suspended in a small amount of THF (distilled from Na, benzophenone). A small amount of a solution of the bromobenzene derivative in THF was added to the magnesium. The reaction was started by shortly heating to reflux. In order to keep the solution at reflux, further bromobenzene solution was added. After complete addition of the bromobenzene, the mixture was heated to reflux for 2 min to ensure complete reaction. Then the Grignard solution was cooled to 0 °C and a solution of the aldehyde in THF was added during 15 min. After stirring at room temperature for at least 2 h, the solution was hydrolyzed with a 0.5 M aqueous NH$_4$Cl-solution and extracted three times with Et$_2$O. The combined organic phases were washed with water, dried with MgSO$_4$, and filtered. Evaporation of the solvent under vacuum gave the crude product. In many cases, the benzhydrols were contaminated by the corresponding benzophenones. To

1. Kinetics of the Solvolyses of Fluoro-Substituted Benzhydryl Derivatives

remove this by-product, the product mixture was treated with sodium borohydride in ethanol for 12 h at ambient temperature. The solution was then hydrolyzed with 0.5 M aqueous NH$_4$Cl-solution and extracted with dichloromethane. The solution was dried with MgSO$_4$ and filtered. Evaporation of the solvent in vacuo gave the crude benzhydrol, which was distilled in vacuo to give the benzhydrol as a colorless oil.

3-Fluorobenzhydrol (7-OH) was obtained from magnesium (4.06 g, 167 mmol), bromobenzene (25.0 g, 159 mmol) in THF (20 mL) and 3-fluorobenzaldehyde (19.7 g, 159 mmol) in THF (30 mL). The reduction of the additionally generated benzophenone was carried out with NaBH$_4$ (1.16 g, 30.7 mmol) in ethanol (50 mL). The crude product was distilled under vacuum (123 °C/3.6 ×10^{-2} mbar) to give a colorless oil (25.6 g, 80 %). Spectroscopic data are in agreement with previously published data.[13]

^{1}H NMR (400 MHz, CDCl$_3$): δ = 2.65 (d, $^3J_{HH}$ = 3.3 Hz, 1 H, OH), 5.69 (d, $^3J_{HH}$ = 3.0 Hz, 1 H, C$\underline{H}$OH), 6.86-6.95 (m, 1 H, 4-Ar$\underline{H}$), 7.02-7.10 (m, 2 H, 2,6-ArH), 7.19-7.31 ppm. (m, 6 H, ArH);

^{13}C NMR {^{1}H}(75.5 MHz, CDCl$_3$): δ = 75.6 (d, J_{CF} = 1.9 Hz, CHOH), 113.4 (d, $^2J_{CF}$ = 22.2 Hz, 2-Ar), 114.3 (d, J_{CF} = 21.2 Hz, 4-Ar), 122.0 (d, J_{CF} = 2.9 Hz, 6-Ar), 126.6 (s, 2',6'-Ar), 127.8 (s, 4'-Ar), 128.6 (s, 2 C, 3',5'-Ar), 129.9 (d, J_{CF} = 8.2 Hz, 5-Ar), 143.3 (s, 1'-Ar), 146.3 (d, J_{CF} = 6.7 Hz, 1-Ar), 162.9 ppm. (d, $^1J_{CF}$ = 245.9 Hz, 3-Ar);

^{19}F NMR (282 MHz, CDCl$_3$): δ = -112.7 ppm. (ddd, $^3J_{FH}$ = 10.3 Hz, $^3J_{FH}$ = 8.7 Hz, $^4J_{FH}$ = 5.7 Hz, 3,3'-F);

MS (+EI): m/z (%) = 202.1 (28) [M$^+$], 201.1 (15) [M$^+$–H], 183.1 (12) [M–F], 123.0 (43) [C$_7$H$_4$FO$^+$], 105.0 (100) [C$_7$H$_5$O$^+$], 97.0 (11) [C$_6$H$_6$F$^+$], 96.0 (11) [C$_6$H$_5$F$^+$], 95.0 (14) [C$_6$H$_4$F$^+$], 79.0 (25) [C$_6$H$_7^+$], 78.0 (20) [C$_6$H$_6^+$], 77.0 (23) [C$_6$H$_5^+$];

HR-MS (+EI) found: 202.0792 calcd. for C$_{13}$H$_{11}$FO: 202.0794.

3,3'-Difluorobenzhydrol (4-OH) was obtained from magnesium (3.65 g, 150 mmol), 1-bromo-3-fluorobenzene (25.0 g, 143 mmol) in THF (20 mL) and 3-fluorobenzaldehyde

1. Kinetics of the Solvolyses of Fluoro-Substituted Benzhydryl Derivatives

(17.7 g, 143 mmol) in THF (30 mL). The reduction of the additionally generated benzophenone was carried out with NaBH$_4$ (1.16 g, 30.7 mmol) in ethanol (50 mL). The crude product was distilled (106 °C/5 ×10^{-3} mbar) in the vacuum to give a colorless oil (24.8 g, 79 %).

^{1}H NMR (300 MHz, CDCl$_3$): δ = 2.89 (d, $^3J_{HH}$ = 3.5 Hz, 1 H, OH), 5.67 (d, $^3J_{HH}$ = 3.2 Hz, 1 H, C$\underline{H}$OH), 6.89-6.96 (m, 2 H, 4-ArH), 6.99-7.06 (m, 4 H, ArH), 7.21-7.28 ppm. (m, 2 H, ArH);

^{13}C NMR {^{1}H}(75.5 MHz, CDCl$_3$): δ = 75.0 (t, $^3J_{CF}$ = 1.9 Hz, CHOH), 113.4 (d, $^2J_{CF}$ = 22.6 Hz, 2-Ar), 114.7 (d, $^2J_{CF}$ = 21.2 Hz, 4-Ar), 122.1 (d, J_{CF} = 2.9 Hz, 6-Ar), 130.1 (d, J_{CF} = 8.2 Hz, 5-Ar), 145.7 (d, J_{CF} = 6.7 Hz, 1-Ar), 162.9 ppm. (d, $^1J_{CF}$ = 246.5 Hz, 3-Ar);

^{19}F NMR (282 MHz, CDCl$_3$): δ = -112.3 ppm. (ddd, $^3J_{FH}$ = 9.5 Hz, $^3J_{FH}$ = 8.7 Hz, $^4J_{FH}$ = 5.7 Hz, 3-F);

MS (+EI): m/z (%) = 221.1 (14) [M$^+$+H], 220.1 (100) [M$^+$], 219.1(38) [M$^+$–H], 203.1 (15) [M$^+$-OH], 202.1 (14) [M$^+$–F+H] 201.1 (39) [M$^+$–F], 183.1 (12), 125.0 (14) [M–C$_6$H$_4$F], 123.0 (80) [C$_7$H$_4$FO$^+$], 97.0 (15) [C$_6$H$_6$F$^+$], 96.0 (12) [C$_6$H$_5$F$^+$], 95.0 (13) [C$_6$H$_4$F$^+$];

HR-MS (+EI) found: 220.0697 calcd. for C$_{13}$H$_{10}$F$_2$O: 220.0700.

3,5-Difluorobenzhydrol (5-OH) was obtained from magnesium (0.79 g, 33 mmol), 1-bromo-3,5-difluorobenzene (6.0 g, 31 mmol) in THF (10 mL) and benzaldehyde (3.3 g, 31 mmol) in THF (10 mL). The reduction of the additionally generated benzophenone was carried out with NaBH$_4$ (1.1 g, 29 mmol) in ethanol (20 mL). The crude product was distilled (170 °C/1.9 ×10^{-2} mbar) under vacuum to give a colorless oil (4.8 g, 70 %).

^{1}H NMR (300 MHz, CDCl$_3$): δ = 2.50 (s, 1 H, OH), 5.72 (s, 1 H, C$\underline{H}$OH), 6.67 (tt, $^3J_{HF}$ = 8.9 Hz, $^4J_{HH}$ = 2.4 Hz 1 H, 4-ArH), 6.85-6.93 (m, 2 H, 2,6-ArH), 7.25-7.38 ppm. (m, 5 H, ArH);

^{13}C NMR {^{1}H}(75.5 MHz, CDCl$_3$): δ = 75.4 (t, $^3J_{CF}$ = 2.2 Hz, 1 C, $\underline{C}$HOH), 102.7 (t, $^2J_{CF}$ = 25.4 Hz, 1 C, 4-Ar), 109.0–109.4 (m, AXX'-system, 2 C, 2,6-Ar), 126.6 (s, 2 C,

1. Kinetics of the Solvolyses of Fluoro-Substituted Benzhydryl Derivatives

2',6'-Ar), 128.2 (s, 1 C, 4'-Ar), 128.8 (s, 2 C, 3',5'-Ar), 142.7 (s, 1 C, 1'-Ar), 147.7 (d, J_{CF} = 8.4 Hz, 2 C, 1-Ar), 163.0 ppm. (dd, $^1J_{CF}$ = 248.5 Hz, $^3J_{CF}$ = 12.6, 2 C, 3,5-Ar); ^{19}F NMR (282 MHz, CDCl$_3$): −109.5 - −109.4 ppm. (m, 2 F, 3,5-F); MS (+ EI): m/z (%) = 220.1 (43) [M$^+$], 219.1 (49) [M$^+$−H], 204.1 (11), 203.1 (9) [M$^+$−OH], 201.1 (22) [M$^+$−F],141.0 (55) [C$_7$H$_3$F$_2$O], 113.0 (18) [C$_6$H$_3$F$_2^+$], 108.0 (22), 107.0 (36) [C$_7$H$_7$O$^+$], 106.0 (12), 105.0 [C$_7$H$_5$O$^+$], 97.1 (11) [C$_6$H$_6$F$^+$], 83.1 (12), 79.0 (55) [C$_6$H$_7^+$], 78.0 (16) [C$_6$H$_6^+$], 77.0 (50) [C$_6$H$_5^+$], 69.0 (13), 57.1 (16), 51.0 (14) [C$_4$H$_3^+$], 44.0 (16), 43.0 (11);

HR-MS (+EI) found: 220.0691 calcd. for C$_{13}$H$_{10}$F$_2$O: 220.0700

Elemental Analysis: Calculated for C$_{13}$H$_{10}$F$_2$O: C, 70.90; H, 4.58.
                      Found: C, 70.56; H, 4.82.

3,3',5-Trifluorobenzhydrol (3-OH) was obtained from magnesium (3.65 g, 150 mmol), 1-bromo-3,5-difluorbenzene (27.6 g, 143 mmol) in THF (20 mL) and 3-fluorobenzaldehyde (17.7 g, 143 mmol) in THF (30 mL). The reduction of the additionally generated benzophenone was carried out with NaBH$_4$ (1.18 g, 31 mmol) in ethanol (50 mL). The crude product was distilled in the vacuum (130 °C/1.7 ×10^{-2} mbar) to give a colorless oil (21.0 g, 62 %).

^{1}H NMR (300 MHz, CDCl$_3$): δ = 2.88 (s, 1 H, OH), 5.68 (s, 1 H, C$\underline{H}$OH), 6.68 (tt, $^3J_{HF}$ = 8.9 Hz, $^4J_{HH}$ = 2.4 Hz 1 H, 4-ArH), 6.81-6.89 (m, 2 H, 2,6-ArH), 6.93-7.09 (m, 3 H, ArH), 7.24-7.33 ppm. (m, 1 H, 5'-ArH);

^{13}C NMR {^{1}H}(75.5 MHz, CDCl$_3$): δ= 74.7 (td, $^4J_{CF}$ = 2.3 Hz, $^4J_{CF}$ = 2.3 Hz, CHOH), 103.1 (t, $^2J_{CF}$ = 25.4 Hz, 1 C, 4-Ar), 109.1–109.4 (m, AXX'-system, 2,6-Ar), 113.5 (d, $^2J_{CF}$ = 22.2 Hz, 2'-Ar), 115.1 (d, $^2J_{CF}$ = 21.2 Hz, 4'-Ar), 122.1 (d, $^4J_{CF}$ = 3.0 Hz, 6'-Ar), 130.4 (d, $^3J_{CF}$ = 8.2 Hz, 5'-Ar), 145.1 (d, $^3J_{CF}$ = 6.7 Hz, 1'-Ar), 147.0 (t, $^3J_{CF}$ = 8.4 Hz, 1-Ar), 163.0 (d, $^1J_{CF}$ = 246.9 Hz, 3'-Ar), 163.1 ppm. (dd, $^1J_{CF}$ = 249.1 Hz, $^3J_{CF}$ = 12.6 Hz, 3,5-Ar);

1. Kinetics of the Solvolyses of Fluoro-Substituted Benzhydryl Derivatives

^{19}F NMR (282 MHz, CDCl$_3$): -109.0 - -108.9 (m, 3,5-F); -111.9 - -112.0 ppm. (m, 3'-F); MS (+EI): m/z (%) = 239.1 (5) [M$^+$+H], 238.1 (32) [M$^+$], 237.1 (10) [M$^+$-H], 219.0 (19) [M$^+$-F], 143.4 (10) [C$_7$H$_5$F$_2$O$^+$], 141.3 (61) [C$_7$H$_3$F$_2$O$^+$], 125.2 (16) [C$_7$H$_6$FO$^+$], 124.2 (14), 123.0 (100) [C$_7$H$_4$FO$^+$], 115.1 (23), 114.1 (18) [C$_6$H$_4$F$_2$$^+$], 113.1 (14) [C$_6H_3F_2$$^+$], 97.0 (31) [C$_6H_6F^+$], 96.0 (31) [C$_6H_5F^+$], 95.0 (24) [C$_6H_4F^+$];
HR-MS (+EI) found: 238.0597 calcd. for C$_{13}$H$_9$F$_3$O: 238.0606;
Elemental Analysis: Calculated for C$_{13}$H$_9$F$_3$O: C, 65.55; H, 3.81.
                    Found: C, 65.25; H, 3.74.

3,3',5,5'-Tetrafluorobenzhydrol (1-OH)
In a flame dried, nitrogen-flushed three-necked round-bottom flask, equipped with a reflux condenser, and two dropping funnels, magnesium (2.64 g, 109 mmol), which was activated with a small amount (tip of a spatula) of iodine at 67 °C, was suspended in a small amount of THF (distilled from Na, benzophenone). A small amount of a solution of 1-bromo-3,5-difluorbenzene (20.0 g, 104 mmol) in THF (30 mL) was added to the magnesium. The reaction was started by short heating to reflux. In order to keep the solution at reflux further bromobenzene solution was added. After complete addition of the bromobenzene the mixture was heated to reflux for 2 minutes, to ensure complete reaction. Then the Grignard solution was cooled to 0 °C and ethyl formate 3.1 g, 42 mmol) in THF (30 mL) was added during 15 min. After stirring at room temperature for at least 2 h, the solution was poured on 0.5 M aqueous NH$_4$Cl-solution (150 mL) and extracted with Et$_2$O (3 × 100 mL). The combined organic phases were washed with water (100 mL), dried with MgSO$_4$, and filtered. Evaporation of the solvent in the vacuum gave the crude product. The crude product (9.31 g, 86 %) was obtained as a pale yellowish, low-melting solid. For synthesis the purity of the crude product is sufficient enough. The product can be further purified by vacuum distillation (103–105 °C/1.0 ×10^{-3} mbar), but care has to be taken that the product does not solidify and clog the condenser.
Mp: 47-50 °C;

1. Kinetics of the Solvolyses of Fluoro-Substituted Benzhydryl Derivatives

^{1}H NMR (300 MHz, CDCl$_3$): δ = 2.84 (br. s, 1 H, OH), 5.68 (s, 1 H, C$\underline{H}$OH), 6.72 (tt, $^3J_{HF}$ = 8.8 Hz, $^4J_{HH}$ = 2.3 Hz, 2 H, 4-H), 6.82–6.89 (m, 4 H, 2,6-H). ^{13}C NMR {^{1}H} (75.5 MHz, CDCl$_3$): δ = 74.4 (quint., J_{CF} = 2.2 Hz, 1 C, $\underline{C}$HOH), 103.5 (t, $^2J_{CF}$ = 25.4 Hz, 2 C, 4-Ar), 109.2–109.5 (m, AXX'-system, 4 C, 2,6-Ar), 146.5 (t, $^3J_{CF}$ = 8.4 Hz, 2 C, 1-Ar), 163.2 ppm. (dd, $^1J_{CF}$ = 249.6 Hz, $^3J_{CF}$ = 12.5 Hz, 4 H, 3,5-Ar);

^{19}F NMR (282 MHz, CDCl$_3$): δ = −108.6 − −108.5 ppm. (m, 4 F, 3,5-F);

MS (+EI): m/z (%) = 256.0 (44) [M$^+$], 255.1 (6) [M$^+$−H], 237.0 (50) [M$^+$−F], 219.0 (32) [M$^+$−2F], 207.0 (13), 206.0 (19), 184.1 (11), 143.0 [C$_7$H$_5$F$_2$O$^+$] (100), 142.0 (28), 141.0 (65), 140.0 (10), 127.0 (15), 115.0 (58) [C$_6$H$_5$F$_2^+$], 114.0 (19) [C$_6$H$_4$F$_2^+$], 95.0 (29) [C$_6$H$_4$F$^+$], 58.0 (25), 44.0 (16), 43.1 (47), 42.1 (24), 41.1 (16);

HR-MS (+EI) found: 256.0358 calcd. for (C$_{13}$H$_8$F$_4$O): 256.0511;

Elemental Analysis: Calculated for C$_{13}$H$_9$F$_3$O: C, 60.95; H, 3.15.
                       Found: C, 60.68; H, 3.09.

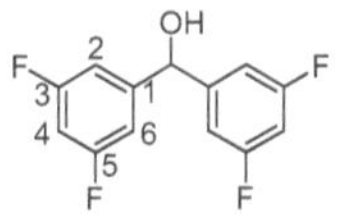

Preparation of the benzhydryl chlorides (**1,3,4,7**)-Cl; general procedure:

The substituted benzhydrol was dissolved in CH$_2$Cl$_2$ at 0 °C. A solution of SOCl$_2$ in CH$_2$Cl$_2$ was added dropwise to the well-stirred solution. After 2 h, the solvent and the remaining SOCl$_2$ was evaporated under vacuum. The remaining crude product was distilled under vacuum to yield the benzhydryl chloride as a colorless oil. All benzhydryl chlorides **1-4**-Cl were prepared by this method. But only (**1-2s**)-Cl were used as the higher substituted benzhydryl chlorides did not solvolyze fast enough to be measured in a reasonable time.

3-Fluorobenzhydryl chloride (7-Cl) was obtained from thionyl chloride (5.1 mL, 70 mmol) in dichloromethane (10 mL) and 3-fluorobenzhydrol (**7-OH**) (10.0 g, 49.5 mmol) in dichloromethane (40 mL). The crude product was distilled in the vacuum (165–170 °C /2.5 ×10^{-3} mbar) to give a colorless oil (8.9 g, 81 %).

1. Kinetics of the Solvolyses of Fluoro-Substituted Benzhydryl Derivatives

^{1}H NMR (400 MHz, CDCl$_3$): δ = 6.08 (s, 1 H, CHCl), 6.95-7.00 (m, 1 H, 4-ArH), 7.12-7.19 (m, 2 H, ArH), 7.27-7.41 ppm. (m, 6 H, ArH);
^{13}C NMR {^{1}H}(101 MHz, CDCl$_3$): δ = 63.3 (d, $^4J_{CF}$ = 1.9 Hz, CHCl), 114.9 (d, $^2J_{CF}$ = 22.8 Hz, 2-Ar or 4-Ar), 115.1 (d, $^2J_{CF}$ = 22.2 Hz, 2-Ar or 4-Ar), 123.4 (d, $^4J_{CF}$ = 3.0 Hz, 6-Ar), 127.7 (s, 2',6',-Ar), 128.3 (s, 4'-Ar), 128.7 (s, 3',5'-Ar), 130.0 (d, $^3J_{CF}$ = 8.2 Hz, 5-Ar), 140.5 (s, 1'-Ar), 143.5 (d, $^3J_{CF}$ = 7.2 Hz, 1-Ar), 162.7 ppm. (d, $^1J_{CF}$ = 246.6 Hz, 3-Ar);
^{19}F NMR (376 MHz, CDCl$_3$): δ = -112.4– -112.3 ppm. (m, 1 F, 3-F);
MS (+EI): m/z (%) = 220.0 (3) [M$^+$], 186.1 (18), 185.1 (100) [M$^+$–Cl], 184.1 (11), 183.1 (31), 165.1 (20);
HR-MS (+EI) found: 220.0443 calcd. for (C$_{13}$H$_{10}$^{35}ClF): 220.0455;
Elemental Analysis: Calculated for C$_{13}$H$_{10}$ClF: C, 70.76; H, 4.57.
                Found: C, 70.63; H, 4.34.

3,3'-Difluorobenzhydryl chloride (4-Cl) was obtained from thionyl chloride (1.4 mL, 19 mmol) and 3,3'-difluorobenzhydrol (4-OH) (3.00 g, 13.6 mmol) in dichloromethane (10 mL). The crude product was distilled in the vacuum (170–174 °C/6.0 ×10^{-3} mbar) to give a colorless oil (2.5 g, 77 %).
^{1}H NMR (300 MHz, CDCl$_3$): δ = 6.05 (s, 1 H, CHCl), 6.97-7.04 (m, 2 H, 4-ArH), 7.10-7.18 (m, 4 H, ArH), 7.29-7.36 (m, 2 H, ArH);
^{13}C NMR {^{1}H}(75.5 MHz, CDCl$_3$): δ = 62.4 (t, $^4J_{CF}$ = 2.0 Hz, CHCl), 114.9 (d, $^2J_{CF}$ = 22.9 Hz, 2-Ar), 115.3 (d, $^2J_{CF}$ = 21.2 Hz, 4-Ar), 123.3 (d, $^4J_{CF}$ = 3.0 Hz, 6-Ar), 130.2 (d, $^3J_{CF}$ = 8.3 Hz, 5-Ar), 142.9 (d, $^3J_{CF}$ = 7.2 Hz, 1-Ar), 162.7 (d, $^1J_{CF}$ = 247.1 Hz, 3-Ar);
^{19}F NMR (282 MHz, CDCl$_3$): δ = -112.0 (m, 3,3'-F);
MS (+EI): m/z (%) = 238.1 (4) [M$^+$], 204.2 (16), 203.2 (100) [M–Cl'], 202.1 (12), 201.1 (35), 183.1 (28);

1. Kinetics of the Solvolyses of Fluoro-Substituted Benzhydryl Derivatives

Elemental Analysis: Calculated for $C_{13}H_9ClF_2$: C, 65.42; H, 3.80.
Found: C, 65.46; H, 3.68.

3,3',5-Trifluorobenzhydryl chloride (3-Cl) was obtained from thionyl chloride (1.3 mL, 18 mmol) and 3,3',5-trifluorobenzhydrol (**3-OH**) (3.00 g, 12.6 mmol) in dichloromethane (10 mL). The crude product was distilled in the vacuum (196–198 °C/1.1 ×10^{-2} mbar) to give a colorless oil (2.3 g, 72 %).

^{1}H NMR (300 MHz, CDCl$_3$): δ = 5.98 (s, 1 H, CHCl), 6.74 (tt, $^3J_{HF}$ = 8.7 Hz, $^4J_{HH}$ = 2.3 Hz ,1 H, 4-ArH), 6.89-7.04 (m, 3 H, ArH), 7.08-7.16 (m, 2 H, ArH), 7.28-7.35 ppm. (m, 1 H, ArH);

^{13}C NMR {^{1}H}(75.5 MHz, CDCl$_3$): δ= 61.8 (td, $^4J_{CF}$ = 2.3 Hz, $^4J_{CF}$ = 2.1 Hz, CHCl), 103.8 (t, $^2J_{CF}$ = 25.3 Hz, 4-Ar), 110.7–111.1 (m, AXX'-system, 2,6-Ar), 114.9 (d, $^2J_{CF}$ = 23.0 Hz, 2'-Ar or 4'-Ar), 115.7 (d, $^2J_{CF}$ = 21.2 Hz, 2'-Ar or 4'-Ar), 123.3 (d, $^4J_{CF}$ = 3.0 Hz, 6'-Ar), 130.4 (d, $^3J_{CF}$ = 8.3 Hz, 5'-Ar), 142.3 (d, $^3J_{CF}$ = 7.2 Hz, 1'-Ar), 144.2 (t, $^3J_{CF}$ = 9.0 Hz, 1-Ar), 162.8 (d, $^1J_{CF}$ = 247.5 Hz, 3'-Ar), 163.0 ppm. (dd, $^1J_{CF}$ = 249.4 Hz, $^3J_{CF}$ = 12.1 Hz, 3,5-Ar);

^{19}F-NMR (282 MHz, CDCl$_3$): −108.4 - −108.5 ppm. (m, 3,5-F); −111.6 (m, 3'-F);

MS (+EI): m/z (%) = 256.1 (3) [M$^+$], 237.2 (4) 222.2 (36), 221.1 (100) [M−Cl$^-$], 219.1 (34), 201.1 (35);

Elemental Analysis: Calculated for $C_{13}H_8ClF_3$: C, 60.84; H, 3.14.
Found: C, 60.61; H, 3.16.

3,3',5,5'-Tetrafluorobenzhydryl chloride (1-Cl) was obtained from thionyl chloride (1.2 mL, 18 mmol) and 3,3',5,5'-tetrafluorobenzhydrol (**1-OH**) (3.00 g, 11.7 mmol) in

1. Kinetics of the Solvolyses of Fluoro-Substituted Benzhydryl Derivatives

dichloromethane (10 mL). The crude product was distilled in the vacuum (196–198 °C/5.0 ×10^{-3} mbar) to give a colorless oil (2.6 g, 82 %).
^{1}H NMR (300.10 MHz, CDCl$_3$): δ = 5.94 (s, 1 H, CHCl), 6.78 (tt, $^3J_{HF}$ = 8.7 Hz, $^4J_{HH}$ = 2.3 Hz, 2 H, 4-H), 6.88-6.96 ppm. (m, 4 H, 2,5-H);
^{13}C NMR {^{1}H}(75.5 MHz, CDCl$_3$): δ = 61.2 (quint., J_{CF} = 2.3 Hz, CHBr),104.2 (t, $^2J_{CF}$ = 25.3 Hz, 4-Ar), 110.7–111.1 (m, AXX'-system, 2,6,-Ar), 143.5 (t, $^3J_{CF}$ = 9.1 Hz, 1-Ar), 163.1 ppm. (dd, $^1J_{CF}$ = 250.6 Hz, $^3J_{CF}$ = 12.8 Hz, 3,5-Ar);
^{19}F-NMR (282 MHz, CDCl$_3$): −108.1 - −108.0 ppm. (m, 3,5-F);
MS (+EI): m/z (%) = 274.1 (4) [M$^+$], 240.2 (65.12), 239.2 [M−Cl$^-$], 237.1 (19), 219.1 (40.3);
Elemental Analysis: Calculated for C$_{13}$H$_7$ClF$_4$: C, 56.85; H, 2.57.
  Found: C, 56.75; H,2.53.

Method for the preparation of the benzhydryl bromides (**1,3,4,5,7**)-Br; general procedure:
In a flame dried, nitrogen-flushed Schlenk flask, equipped with a dropping funnel a solution of the substituted benzhydrol in dichloromethane was prepared at 0 °C. Phosphorus tribromide in dichloromethane was added to the well stirred solution. After addition of PBr$_3$ the cooling bath was removed and the solution was allowed to reach room temperature. Stirring was continued for at least three hours; then the solvent and PBr$_3$ were evaporated in the vacuum. Distillation under vacuum yielded the benzhydryl bromides as colorless oil.

3-Fluorobenzhydryl bromide (7-Br) was obtained from phosphorus tribromide (8.1 g, 30 mmol), 3-fluorobenzhydrol (**7-OH**) (5.0 g, 25 mmol) in dichloromethane (35 mL). The crude product was distilled in the vacuum (175-180 °C/1.0 ×10^{-3} mbar) to give a colorless oil (5.2 g, 80 %).

1. Kinetics of the Solvolyses of Fluoro-Substituted Benzhydryl Derivatives

^{1}H NMR (300 MHz, CDCl$_3$): δ = 6.22 (s, 1 H, C$\underline{H}$Br), 6.92-6.99 (m, 1 H, 4-ArH), 7.15-7.45 ppm. (m, 8 H, ArH);
^{13}C NMR {^{1}H}(75.5 MHz, CDCl$_3$): δ = 54.0 (d, $^4J_{CF}$ = 2.0 Hz, CHBr), 115.0 (d, $^2J_{CF}$ = 21.2 Hz, 4-Ar), 115.6 (d, $^2J_{CF}$ = 22.9 Hz, 2-Ar), 124.0 (d, $^4J_{CF}$ = 3.0 Hz, 6-Ar), 128.3 (s, 2',4',6'-Ar), 128.7 (s, 3',5'-Ar),130.0 (d, $^3J_{CF}$ = 8.3 Hz, 5-Ar), 140.4 (s, 1'-Ar), 143.5 (d, $^3J_{CF}$ = 7.3 Hz, 1-Ar), 162.6 ppm, (d, $^1J_{CF}$ = 246.8 Hz, 3-Ar),
^{19}F NMR (282 MHz, CDCl$_3$): δ = −112.3 - −112.2 ppm. (ddd, $^3J_{FH}$ = 9.6 Hz, $^3J_{FH}$ = 8.2 Hz, $^4J_{FH}$ = 5.6 Hz, 3-F);
MS (+EI): m/z (%) = 369.2 (26) [2M$^+$−2Br$^-$+H], 368.2 (100) [2M$^+$−2Br$^-$], 308.2 (22), 271.2 (20), 269.2 (14), 265.2 (55) [M$^+$ + H], 264.1 (10) [M$^+$] ,263.1 (58) [M$^+$- H], 262.2 (11), 261.2 (30), 259.2 (12);
Elemental Analysis: Calculated for C$_{13}$H$_{10}$BrF$_1$: C, 58.89; H, 3.80.
Found: C, 58.79; H, 3.81.

3,3'-Difluorobenzhydryl bromide (4-Br) was obtained from phosphorus tribromide (13.5 g, 49.9 mmol), 3,3'-difluorobenzhydrol (4-OH) (10.0 g, 45.4 mmol) in dichloromethane (35 mL).The crude product was distilled in the vacuum (175–180 °C/1.0×10^{-3} mbar) to give a colorless oil (9.0 g, 70 %).
^{1}H NMR (300 MHz, CDCl$_3$): δ = 6.18 (s, 1 H, CHBr), 6.96-7.02 (m, 1 H, ArH), 7.14-7.35 ppm. (m, 8 H, ArH);
^{13}C NMR {^{1}H}(75.5 MHz, CDCl$_3$): δ = 52.7 (t, $^3J_{CF}$ = 2.0 Hz, CHBr), 115.3 (d, $^2J_{CF}$ = 21.2 Hz, 4-Ar), 115.6 (d, $^2J_{CF}$ = 22.9 Hz, 2-Ar), 124.0 (d, $^4J_{CF}$ = 3.0 Hz, 6-Ar), 130.2 (d, $^3J_{CF}$ = 8.3 Hz, 5-Ar), 142.9 (d, $^4J_{CF}$ = 7.2 Hz, 1-Ar), 162.6 ppm. (d, $^1J_{CF}$ = 247.0 Hz, 3-Ar);
^{19}F NMR (282.37 MHz, CDCl$_3$): δ = -112.0 ppm. (ddd, $^3J_{FH}$ = 9.7 Hz, $^3J_{FH}$ = 8.5 Hz, $^4J_{FH}$ = 5.9 Hz, 3,3'-F);
MS (+EI): m/z (%) = 281.2 (<1) [M$^+$−H], 204.2 (32), 203.0 (100) [M−Br$^-$], 202.1 (21), 201.2 (48), 183.2 (40), 123.1 (12) [C$_7$H$_4$FO$^+$], 58.1 (36), 43.1 (66);

1. Kinetics of the Solvolyses of Fluoro-Substituted Benzhydryl Derivatives

Elemental Analysis: Calculated for $C_{13}H_9BrF_2$: C, 55.15; H, 3.20.
Found: C, 55.09; H, 3.15.

3,5-Difluorbenzhydryl bromide (5-Br) was obtained from phosphorus tribromide (1.60 g, 5.9 mmol), 3,5-difluorobenzhydrol (5-OH) (1.00 g, 4.5 mmol) in dichloromethane (5 mL). The crude product was distilled in the vacuum (170-175 °C/2.0 ×10^{-3} mbar) to give a colorless oil (0.87 g, 68 %).

^{1}H NMR (300 MHz, CDCl$_3$): δ = 6.14 (s, 1 H, CHBr), 6.70 (tt, $^3J_{HF}$ = 8.7 Hz, $^4J_{HH}$ = 2.3 Hz ,1 H, 4-ArH), 6.93-7.01 (m, 2 H, ArH), .7.26-7.43 ppm. (m, 5 H, ArH);
^{13}C NMR {^{1}H}(75.5 MHz, CDCl$_3$): δ = 53.1 (t, $^3J_{CF}$ = 2.2 Hz, CHBr), 103.5 (t, $^2J_{CF}$ = 25.3 Hz, 4-Ar), 111.5-111.8 (m, AXX'-system, 2,6-Ar), 128.3 (s, 2',6'-Ar), 128.6 (s, 2 C, 4'-Ar), 128.8 (s, 3',5'-Ar), 139.9 (s, 1'-Ar), 144.8 (t, $^3J_{CF}$ = 9.0 Hz, 1-Ar), 162.8 ppm. (dd, $^1J_{CF}$ = 249.3 Hz, $^3J_{CF}$ = 12.8 Hz, 2 C, 3,5-Ar);
^{19}F-NMR (282 MHz, CDCl$_3$): −108.8 - −108.9 ppm. (m, 3,5-F);
MS (+EI): m/z (%) = 204.1 (29), 203.1 (100) [M$^+$−Br], 202.1 (19), 201.1 (48), 184.1 (10), 183.1 (53), 141.1 (10) [C$_7$H$_3$F$_2$O$^+$], 125.2 (11), 123.2 (10), 111.2 (15), 109.2 (10), 105.1 (24) [C$_7$H$_5$O$^+$], 97.2 (20), 95.1 (16), 91.2 (14), 85.2 (14), 83.1 (17), 81.1 (18), 77.1 (10) [C$_6$H$_5^+$], 71.1 (16), 69.1 (19), 57.1 (25), 55.1 (19), 44.0 (37), 43.1 (20), 41.1 (26);
Elemental Analysis: Calculated for $C_{13}H_9BrF_2$: C, 55.15; H, 3.20.
Found: C, 54.79; H, 3.16.

3,3',5-Trifluorobenzhydryl bromide (3-Br) was obtained from phosphorus tribromide (13.6 g, 50.2 mmol), 3,3',5-trifluorobenzhydrol (3-OH) (10.0 g, 42.0 mmol) in dichloromethane (35 mL). The crude product was distilled in the vacuum (185-188 °C/1.0 ×10^{-3} mbar) to give a colorless oil (10.1 g, 80 %).

1. Kinetics of the Solvolyses of Fluoro-Substituted Benzhydryl Derivatives

^{1}H NMR (300 MHz, CDCl$_3$): δ = 6.11 (s, 1 H, CHBr), 6.75 (tt, $^3J_{HF}$ = 8.7 Hz, $^4J_{HH}$ = 2.3 Hz ,1 H, 4-ArH), 6.92-7.04 (m, 3 H, ArH), 7.12-7.20 (m, 2 H, ArH), 7.28-7.36 ppm. (m, 1 H, ArH);

^{13}C NMR {^{1}H}(75.5 MHZ, CDCl$_3$): δ= 51.8 (td, $^4J_{CF}$ = 2.3 Hz, $^4J_{CF}$ = 2.1 Hz, CHBr), 103.8 (t, $^2J_{CF}$ = 25.3 Hz, 4-Ar), 111.4–111.8 (m, AXX'-system, 2,6-Ar), 115.6 (d, $^2J_{CF}$ = 22.3 Hz, 2'-Ar or 4'-Ar), 115.6 (d, $^2J_{CF}$ = 21.2 Hz, 2'-Ar or 4'-Ar), 123.9 (d, $^4J_{CF}$ = 3.0 Hz, 6'-Ar), 130.4 (d, $^3J_{CF}$ = 8.3 Hz, 5'-Ar), 142.2 (d, $^3J_{CF}$ = 7.3 Hz, 1'-Ar), 144.2 (t, $^3J_{CF}$ = 9.1 Hz, 1-Ar), 162.7 (d, $^1J_{CF}$ = 247.5 Hz, 3'-Ar), 162.8 ppm. (dd, $^1J_{CF}$ = 249.7 Hz, $^3J_{CF}$ = 12.8 Hz, 3,5-Ar);

^{19}F-NMR (282 MHz, CDCl$_3$): –108.5 - –108.4 ppm. (m, 3,5-F); -111.6– -111.7 (m, 3'-F);

MS (+EI): m/z (%) = 302.2 (11) [M$^+$], 222.2 (23), 221.2 (100) [M$^+$–Br], 220.2 (18), 219.1 (51), 201.1 (50);

HR-MS (+EI) found: 299.9728 calcd. for C$_{13}$H$_8$^{79}BrF$_3$: 299.9761;

Elemental Analysis: Calculated for C$_{13}$H$_8$BrF$_3$: C, 51.86; H, 2.68.
                      Found: C, 51.76; H, 2.53.

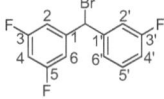

3,3',5,5'-Tetrafluorobenzhydryl bromide (1-Br) was obtained from phosphorus tribromide (3.80 g, 14.0 mmol), 3,3',5,5'-tetrafluorbenzhydrol (1-OH) (3.00 g, 11.7 mmol) in dichloromethane (35 mL). The crude product was distilled in the vacuum (188–193 °C/1.0 ×10^{-3} mbar) to give a colorless oil (2.7 g, 73 %).

^{1}H NMR (300.10 MHz, CDCl$_3$): δ =6.05 (s, 1 H, CHBr), 6.76 (tt, $^3J_{HF}$ = 8.7 Hz, $^4J_{HH}$ = 2.2 Hz, 2 H, 4-H), 6.90-6.99 ppm. (m, 4 H, 2,5-H);

^{13}C NMR {^{1}H}(75.5 MHz, CDCl$_3$): δ = 50.9 (quint., J_{CF} = 2.3 Hz, CHBr),104.1 (t, $^2J_{CF}$ = 25.2 Hz, 4-Ar), 111.4–111.7 (m, AXX'-system, 2,6,-Ar), 143.5 (t, $^3J_{CF}$ = 9.0 Hz, 1-Ar), 162.9 ppm. (dd, $^1J_{CF}$ = 250.0 Hz, $^3J_{CF}$ = 12.7 Hz, 3,5-Ar);

^{19}F-NMR (282 MHz, CDCl$_3$): –108.2 - –108.1 ppm. (m, 3,5-F);

MS (+EI): m/z (%) = 320.1 (18) [M$^+$], 303.2 (16), 302.2 (100) [M–F+H$^+$], 240.1 (12), 239.1 (100), 237.1 (22), 219.1 (34);

41

1. Kinetics of the Solvolyses of Fluoro-Substituted Benzhydryl Derivatives

Elemental Analysis: Calculated for $C_{13}H_7BrF_4$: C, 48.93; H, 2.21.
Found: C, 48.79; H, 2.25.

Method for the preparation of the benzhydryl tosylates (1,3,4,7)-OTs; general procedure:
In a flame dried, nitrogen-flushed, opaque Schlenk flask, a 1.3-fold excess of finely ground silver tosylate (2-4 g) was suspended in dichloromethane. The benzhydryl bromide (1.5 to 4.1 g) was added to the well stirred suspension. After stirring for at least 12 hours, the solvent was removed under reduced pressure. The remaining solid was extracted twice at room temperature with diethylether (10 mL). The remaining silver salts were filtered off. The resulting solution was concentrated by removing part of the diethylether under reduced pressure until colorless crystals started to precipitate. The solution was cooled to –34 °C and the product was allowed to crystallize over night. The resulting colorless crystals of benzhydryl tosylate were filtered, washed with a very small amount of cold ether and dried under high vacuum to give the benzhydryl tosylates in mediocre yields. During the complete synthesis care has to be taken to exclude traces of water.

3-Fluorobenzhydryl tosylate (7-OTs) was obtained from silver tosylate (4.11 g, 14.7 mmol), 3-fluorobenzhydryl bromide (**7-Br**) (3.00 g, 11.3 mmol) in dichloromethane (15 mL) yielding in the product as colorless crystals (1.5 g, 37 %).
Mp: 68.9 - 70.4 °C (decomp.);
^{1}H NMR (300 MHz, CDCl$_3$): δ = 2.37 (s, 3 H, CH$_3$), 6.50 (s, 1 H, CHOTs), 6.88-6.98 (m, 2 H, 2,4-ArH, TolH), 6.99-7.04 (m, 1 H, 6.ArH), 7.14-7.29 (m, 8 H, ArH), 7.60-7.64 ppm. (m, 2 H, TolH);
^{13}C NMR {^{1}H}(75.5 MHz, CDCl$_3$): δ = 21.7 (s, CH$_3$), 83.7 (d, $^4J_{CF}$ = 2.0 Hz, CHOTs), 114.4 (d, $^2J_{CF}$ = 22.9 Hz, 2-Ar), 115.4 (d, $^2J_{CF}$ = 21.2 Hz, 4-Ar), 123.0 (d, $^4J_{CF}$ = 3.0 Hz, 6-Ar), 127.4 (s, 2',6'-Ar), 127.9 (s, TolH), 128.7 (s, 3',5'-Ar), 128.8 (s, 4'-Ar), 129.6 (s, TolH), 130.2 (d, $^3J_{CF}$ = 8.2 Hz, 5-Ar), 134.1 (s, Tol), 137.8 (s, 1'-Ar), 140.9 (d, $^3J_{CF}$ = 7.3 Hz, 1-Ar), 144.7 (s, Tol), 162.8 ppm. (d, $^1J_{CF}$ = 246.9 Hz, 3-Ar);

1. Kinetics of the Solvolyses of Fluoro-Substituted Benzhydryl Derivatives

^{19}F NMR (282.38 MHz, CDCl$_3$): δ = −112.3 - −112.2 ppm. (m, 3-F);
MS (+EI): m/z (%) = 356.2 (0.1) [M$^+$], 201.2 (28) [C$_{13}$H$_{10}$FO$^+$], 186.2 (56), 185.2 (100) [M$^+$−OTs], 183.1 (36), 165.1 (20), 123.1 (16) [C$_7$H$_5$FO], 105.1 (36), 91.2 (20), 77.1 (12);
Elemental Analysis: Calculated for C$_{20}$H$_{17}$FO$_3$S: C, 67.40; H, 4.81; S, 9.00.
Found: C, 67.58; H, 5.15; S, 9.21.

3,3'-Difluorobenzhydryl tosylate (4-OTs) was obtained from silver tosylate (3.84 g, 13.8 mmol), 3,3'-difluorobenzhydryl bromid **(4-Br)** (3.00 g, 10.6 mmol) in dichloromethane (15 mL) yielding in the product as colorless crystals (1.4 g, 35 %).
Mp: 84.6-87.4 °C;
^{1}H NMR (300 MHz, CDCl$_3$): δ = 2.36 (s, 3 H, CH$_3$), 6.46 (s, 1 H, CHOTs), 6.86-7.00 (m, 6 H, ArH, TolH), 7.15-7.30 (m, 4 H, ArH), 7.60-7.63 ppm. (m, 2 H, TolH);
^{13}C NMR {^{1}H}(75.5 MHz, CDCl$_3$): δ = 21.5 (s, 1 C, CH$_3$), 82.6 (t, $^4J_{CF}$ = 1.9 Hz, CHOTs), 114.2 (d, $^2J_{CF}$ = 22.9 Hz, 2 C, 2-Ar), 115.6 (d, $^2J_{CF}$ = 21.1 Hz, 4-Ar), 122.8 (d, $^4J_{CF}$ = 3.0 Hz, 6-Ar),127.8 (s, TolH), 129.6 (s, Tol), 130.2 (d, $^4J_{CF}$ = 8.2 Hz, 5-Ar), 133.6 (s, Tol), 140.1(d, $^4J_{CF}$ = 7.2 Hz, 1-Ar), 145.0 (s, Tol), 162.6 ppm. (d, $^1J_{CF}$ = 247.4 Hz, 3-Ar);
^{19}F NMR (282 MHz, CDCl$_3$): δ = −112.0 - −111.9 ppm. (m, 3-F);
MS (+EI): m/z (%) = 374.1 (0.1) [M$^+$], 219.1 (22) [C$_{13}$H$_9$F$_2$O$^+$], 204.0 (45), 203.0 (100) [M$^+$-OTs], 201.0 (46), 183.0 (32), 91.0 (10);
HR-MS (+EI) found: 374.0781 calcd. for (C$_{20}$H$_{16}$F$_2$O$_3$^{32}S): 374.0788;

1. Kinetics of the Solvolyses of Fluoro-Substituted Benzhydryl Derivatives

Elemental Analysis: Calculated for $C_{20}H_{16}F_2O_3S$: C, 64.16; H, 4.31.
Found: C, 63.88; H, 4.27.

3,3',5-Trifluorobenzhydryl tosylate (3-OTs) was obtained from silver tosylate (3.61 g, 12.9 mmol), 3,3',5-trifluorobenzhydryl bromide (3-Br) (3.00 g, 9.96 mmol) in dichloromethane (15 mL) yielding in the product as colorless crystals (1.5 g, 38 %).
Mp: 95.8-98.4 °C;

^{1}H NMR (599 MHz, CDCl$_3$): δ = 2.38 (s, 3 H, CH$_3$), 6.41 (s, 1 H, CHOTs), 6.68-6.75 (m, 3 H, 2,4,6-ArH,), 6.86 (d, $^3J_{HF}$ = 9.3 Hz, 1 H, ArH), 6.95-6.98 (m, 2 H, ArH), 7.20 (d, $^3J_{HH}$ = 8.0 Hz, 2 H, TolH), 7.23-7.26 (m, 1 H, 5'-ArH), 7.64 ppm. (d, $^3J_{HH}$ = 8.0 Hz, 2 H, TolH);

^{13}C NMR {^{1}H}(150.7 MHz, CDCl$_3$): δ = 21.5 (s, CH$_3$), 81.8 (td, $^4J_{CF}$ = 2.1 Hz, $^4J_{CF}$ = 2.2 Hz, CHOTs), 104.0 (t, $^2J_{CF}$ = 25.2 Hz, 4-Ar), 110.1–110.3 (m, AXX'-system, 2,6-Ar), 114.2 (d, $^2J_{CF}$ = 22.9 Hz, 2'-Ar), 115.9 (d, $^2J_{CF}$ = 21.1 Hz, 4'-Ar), 122.8 (d, $^4J_{CF}$ = 3.1 Hz, 6'-Ar), 127.8 (s, Tol), 129.7 (s, Tol), 130.4 (d, $^4J_{CF}$ = 8.2 Hz, 5'-Ar), 133.6 (s, Tol), 139.5 (d, $^4J_{CF}$ = 7.3 Hz, 1'-Ar), 141.6 (t, $^4J_{CF}$ = 9.0 Hz, 1-Ar), 145.1 (s, Tol), 162.6 (d, $^1J_{CF}$ = 247.7 Hz, 3'-Ar), 162.9 ppm. (dd, $^1J_{CF}$ = 250.1 Hz, $^3J_{CF}$ = 12.6 Hz, 3,5-Ar);

^{19}F-NMR (282 MHz, CDCl$_3$): -111.7 - −111.6 (m, 3'-F), −108.3 - −108.2 (m, 3,5-F);

MS (+EI): m/z (%) = 392.1 (0.1) [M$^+$], 237.1 (29) [C$_{13}$H$_8$F$_3$O$^+$], 222.1 (16), 221.1 (100) [M$^+$–OTs], 220.0 (16), 219.0 (37), 201.1 (22), 200.2 (18);

1. Kinetics of the Solvolyses of Fluoro-Substituted Benzhydryl Derivatives

HR-MS (+EI) found: 392.0688 calcd. for ($C_{20}H_{15}F_3O_3{}^{32}S$): 392.0694;
Elemental Analysis: Calculated for $C_{20}H_{15}F_3O_3S$: C, 61.22; H, 3.85; S, 8.17.
Found: C, 61.00; H, 4.02; S, 7.93.

3,3',5,5'-Tetrafluorobenzhydryl tosylate (1-OTs) was obtained from silver tosylate (2.40 g, 8.60 mmol), 3,3',5,5'-tetrafluorobenzhydryl bromide (1-Br) (1.50 g, 4.70 mmol) in dichloromethane (15 mL) yielding in the product as colorless crystals (0.43 g, 21 %).
Mp: 96.1-98.6 °C
^{1}H NMR (300 MHz, CDCl$_3$): δ = 2.40 (s, 3 H, CH$_3$), 6.36 (s, 1 H, CHOTs), 6.69-6.77 (m, 6 H, ArH,), 7.21-7.25 (m, TolH), 7.63-7.67 ppm. (m, 2 H, TolH);
^{13}C NMR {^{1}H}(75.5 MHz, CDCl$_3$): δ = 21.6 (s, CH$_3$), 81.0 (quint., $^4J_{CF}$ = 2.3 Hz, CHOTs), 104.4 (t, $^2J_{CF}$ = 25.2 Hz, 4-Ar), 110.0–110.4 (m, AXX'-system, 2,6-Ar), 127.9 (s, Tol), 129.7 (s, Tol), 133.3 (s, Tol), 140.9 (t, $^4J_{CF}$ = 9.0 Hz, 1-Ar), 145.4 (s, Tol), 163.0 ppm. (dd, $^1J_{CF}$ = 250.6 Hz, $^3J_{CF}$ = 12.6 Hz, 3,5-Ar);
^{19}F-NMR (282.38 MHz, CDCl$_3$): −108.0 - −107.9 ppm. (m, 3,5-F);
MS (+EI): m/z (%) = 410.1 (0.4) [M$^+$], 255.1 (36) [$C_{13}H_7F_4O^+$], 240.1 (12), 239.0 (100) [M$^+$–OTs], 238.0 (14), 219.0 (29);
HR-MS (+EI) found: 410.0577 calc. ($C_{20}H_{14}F_4O_3{}^{32}S$): 410.0600
Elemental Analysis: Calculated for $C_{20}H_{14}F_4O_3S$: C, 58.53; H, 3.44.
Found: C, 58.31; H, 3.26.

1. Kinetics of the Solvolyses of Fluoro-Substituted Benzhydryl Derivatives

Method for the preparation of the benzhydrylmesylates (**3,4,7**)-OMs:
As described in the main article, it was not possible to isolate the benzhydryl mesylates **3,4,7**-OMs as pure products. For the investigation of their solvolysis reaction diluted reaction mixtures were used. To obtain these mixtures, 0.45 M solutions of the corresponding benzhydryl bromide **3,4,7**-Br in dichloromethane were treated with 1.3 equivalents of silver mesylate. After 12 h, the solution was filtered and diluted to give an approximately 0.2 M solution of the benzhydryl mesylate. The solutions were used immediately for the kinetic experiments.

3,3',5,5'-Tetrafluorobenzhydryl mesylate (1-OMs) was obtained from silver mesylate (0.83 g, 4.1 mmol) and 3,3',5,5'-Tetrafluorobenzhydryl bromide (**1-Br**) (1.0 g, 3.2 mmol) in dichloromethane (10 mL) yielding in the product as colorless crystals (0.8 g, 77 %).
Mp: 59.3-61.4 °C;
^{1}H NMR (300 MHz, CDCl$_3$): δ = 2.94 (s, 3 H, C$\underline{H}_3$), 6.55 (s, 1 H, CHOMs), 6.78-6.86 (m, 2 H, 4,4'-ArH), 6.88-6.95 ppm. (m, 4 H, ArH);
^{13}C NMR {^{1}H}(75.5 MHz, CDCl$_3$): δ = 39.3 (s, CH$_3$), 80.7 (quint., $^4J_{CF}$ = 2.3 Hz, 1 C, CHOMs), 104.8 (t, $^2J_{CF}$ = 25.2 Hz, 4-Ar), 110.1–110.5 (m, AXX'-system, 2 C, 2,6-Ar), 141.2 (t, $^4J_{CF}$ = 8.8 Hz, 1-Ar), 163.3 ppm. (dd, $^1J_{CF}$ = 250.6 Hz, $^3J_{CF}$ = 12.6 Hz, 4 C, 3,5-Ar);
^{19}F-NMR (282 MHz, CDCl$_3$): −107.4 - −107.3 ppm. (m, 3,5-F);
MS (+EI): m/z (%) = 334.0 (15) [M$^+$], 293.2 (20), 256.0 (19), 256.0 (19), 255.1 (98) [C$_{13}$H$_7$F$_4$O$^+$], 240.1 (20), 239.0 (100) [M$^+$ - OTs], 238.0 (44) 237.0 (69), 220.0 (16), 219.0 (83), 149.0 (62), 141.0 (38), 127.1 (11), 113.0 (13) [C$_6$H$_3$F$_2^+$];
HR-MS (+EI) found: 334.0273 calcd. for (C$_{14}$H$_{10}$F$_4$O$_3$^{32}S): 334.0287
Elemental Analysis: Calculated for C$_{14}$H$_{10}$F$_4$O$_3$S: C, 50.30; H, 3.02; S, 9.59.
                Found: C, 50.40; H, 3.13; S, 9.72.

1. Kinetics of the Solvolyses of Fluoro-Substituted Benzhydryl Derivatives

1.6. Kinetics

Hydrolysis or alcoholysis of the benzhydryl derivatives (**1,3,4,5,7**)-X with X = Cl, Br, OMs, OTs led to the formation of the benzhydrols ((**1,3,4,5,7**)-OH) or benzhydryl ethers ((**1,3,4,5,7**)-OR) along with the strong acids HX. The generation of HX resulted in an increase of conductivity. Calibration experiments for two representative systems showed that the initial concentration of benzhydryl bromide or tosylate correlates linearly with the final conductance, in agreement with previous results. Therefore, the solvolysis rate constants can be determined reliably by conductometry. Most reactions were monitored with a conventional conductometer (conductometers: Radiometer Analytical CDM 230 or Tacussel CD 810, Pt electrode: WTW LTA 1/NS). The temperature of the solutions during all kinetic studies was kept constant at 25.0 °C (± 0.1 °C) by using a circulating bath thermostat. For each measurement aliquots of 0.25 mL of a 0.2 M solution of the substrate in dichloromethane were injected to 30 mL of the solvolysis medium. Fast solvolysis reactions, e.g., solvolysis of **7**-OTs in 80 % aqueous acetone, have been measured in a stopped-flow conductometer (Hi-Tech Scientific SF-61 DX2, platinum electrodes, cell volume: 21 µL, cell constant 4.24 cm^{-1}, minimum dead time 2.2 ms) by mixing one equivalent of the benzhydryl derivative in acetone or acetonitrile with 10 equivalents of aqueous acetone or acetonitrile to give solvent mixtures of the desired composition. After injection of the benzhydrylium derivative into the solvolyzing medium an increase of conductance was observed, which was recorded at certain time intervals resulting in about 3000 data points for each measurement. The first-order rate constants k_1 (s^{-1}) were obtained by least squares fitting of the conductance data to a single-exponential equation $G = G_\infty(1-e^{-k_1 t}) + C$. Each rate constant was typically averaged from at least three kinetic runs. In some cases when the solvolysis reaction was very slow only two kinetic runs were recorded. All solvolyses were performed at 25 °C. The following solvents were commercially available with a sufficient quality for the kinetic experiments: Acetone (Acros 99.8 %), acetonitrile (extra dry < 50 ppm), methanol (Acros 99.8 %). Dry ethanol was obtained by distillation of commercially available absolute ethanol from sodium/diethyl phtalate. Dry 2,2,2-trifluorethanol was obtained by distillation of commercially available 2,2,2-trifluorethanol (Apollo Scientific) from Drierite$^©$. As 2,2,2-trifluoroethanol is quite expensive it was recycled by distilling the 2,2,2-trifluoroethanol

1. Kinetics of the Solvolyses of Fluoro-Substituted Benzhydryl Derivatives

with an rotary evaporator and then returning the solvent to the distillation from Drierite[©]. This procedure might have caused the bigger deviation in the individual rate constants of these measurements. Doubly distilled water [Impendance 18.2 Ω] was prepared with a Milli-Q Plus machine from Millipore.

Calibration experiments showed a linear correlation between the initial concentration of the benzhydryl derivatives **7-OTs**, **9-Br** and the conductance at the end of the reaction within the investigated concentration range (Figure 1.3).

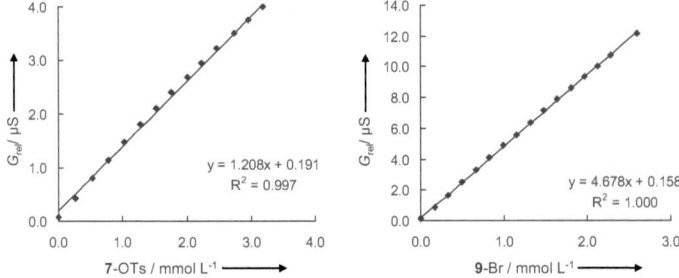

Figure 1.3. Initial concentration of benzhydryl tosylate **7-OTs** and **9-Br** (Ph$_2$CHBr) vs. conductance at t_∞ in 100E and 50A50W respectively. After the addition of a portion of **7-OTs** or **9-Br**, the next conductivity value was taken when the conductivity remained constant for a least 100 s and 200 s respectively.

1. Kinetics of the Solvolyses of Fluoro-Substituted Benzhydryl Derivatives

Table 1.6. Individual rate constants for the solvolysis reactions of **(1,3,4,5,7)**-X.

solvent[a]	benzhydrylium derivative	k_1 (individual)/ s^{-1}	k_1 (average)/s^{-1}
90A10W	7-OTs	1.22×10^{-2}	1.16×10^{-2}
		1.17×10^{-2}	
		1.17×10^{-2}	
		1.13×10^{-2}	
		1.14×10^{-2}	
		1.14×10^{-2}	
	4-OTs	4.47×10^{-4}	4.24×10^{-4}
		4.17×10^{-4}	
		4.08×10^{-4}	
80A20W	7-OTs	5.43×10^{-2}	5.59×10^{-2} [b]
		5.75×10^{-2}	
	4-OTs	2.38×10^{-3}	2.42×10^{-3}
		2.45×10^{-3}	
		2.43×10^{-3}	
	3-OTs	1.11×10^{-4}	1.10×10^{-4}
		1.10×10^{-4}	
80A20W	7-OMs	3.94×10^{-2}	3.87×10^{-2}
		3.68×10^{-2}	
		3.82×10^{-2}	
		4.03×10^{-2}	
	4-OMs	1.38×10^{-3}	1.37×10^{-3}
		1.37×10^{-3}	
		1.35×10^{-3}	
		1.36×10^{-3}	
	3-OMs	5.88×10^{-5}	5.92×10^{-5}
		5.96×10^{-5}	

1. Kinetics of the Solvolyses of Fluoro-Substituted Benzhydryl Derivatives

Table 1.6. (continued)

solvent[a]	benzhydrylium derivative	k_1 (individual)/ s^{-1}	k_1 (average)/s^{-1}
60A40W	7-Br	1.51×10^{-3}	1.47×10^{-3}
		1.45×10^{-3}	
		1.48×10^{-3}	
		1.45×10^{-3}	
50A50A	4-Br	1.55×10^{-4}	1.59×10^{-4}
		1.60×10^{-4}	
		1.60×10^{-4}	
60AN40W	7-OTs	2.14	2.14 [b]
	4-OTs	9.52×10^{-2}	9.53×10^{-2}
		9.95×10^{-2}	
		9.37×10^{-2}	
		9.29×10^{-2}	
	3-OTs	4.35×10^{-3}	4.26×10^{-3}
		4.30×10^{-3}	
		4.24×10^{-3}	
		4.16×10^{-3}	
	1-OTs	1.43×10^{-4}	1.45×10^{-4}
		1.47×10^{-4}	
		1.47×10^{-4}	
60AN40W	4-OMs	4.68×10^{-2}	4.82×10^{-2}
		5.12×10^{-2}	
		4.73×10^{-2}	
		4.77×10^{-2}	
	3-OMs	2.53×10^{-3}	2.52×10^{-3}
		2.45×10^{-3}	
		2.57×10^{-3}	

1. Kinetics of the Solvolyses of Fluoro-Substituted Benzhydryl Derivatives

Table 1.6. (continued)

solvent[a]	benzhydrylium derivative	k_1 (individual)/ s^{-1}	k_1 (average)/s^{-1}
60AN40W	1-OMs	7.89×10^{-5}	8.04×10^{-5}
		8.45×10^{-5}	
		7.79×10^{-5}	
60AN40W	10-Br	4.59	4.57
		4.55	
	9-Br	1.40×10^{-1}	1.44×10^{-1}
		1.48×10^{-1}	
	7-Br	4.39×10^{-3}	4.59×10^{-3}
		4.78×10^{-3}	
		4.61×10^{-3}	
	4-Br	1.22×10^{-4}	1.18×10^{-4}
		1.16×10^{-4}	
		1.16×10^{-4}	
	5-Br	1.60×10^{-4}	1.59×10^{-4}
		1.58×10^{-4}	
60AN40W	7-Cl	2.67×10^{-4}	2.67×10^{-4}
		2.66×10^{-4}	
100E	7-OTs	7.60×10^{-2}	8.22×10^{-2}
		7.60×10^{-2}	
		8.75×10^{-2}	
		8.05×10^{-2}	
		8.75×10^{-2}	
		8.57×10^{-2}	
	4-OTs	3.24×10^{-3}	3.35×10^{-3}
		3.28×10^{-3}	
		3.48×10^{-3}	
		3.38×10^{-3}	

1. Kinetics of the Solvolyses of Fluoro-Substituted Benzhydryl Derivatives

Table 1.6. (continued)

solvent[a]	benzhydrylium derivative	k_1 (individual)/ s^{-1}	k_1 (average)/s^{-1}
100E	3-OTs	1.96×10^{-4}	1.88×10^{-4}
		1.82×10^{-4}	
		1.84×10^{-4}	
100E	7-OMs	4.45×10^{-2}	4.34×10^{-2}
		4.17×10^{-2}	
		4.17×10^{-2}	
		4.59×10^{-2}	
	4-OMs	1.63×10^{-3}	1.70×10^{-3}
		1.76×10^{-3}	
		1.62×10^{-3}	
		1.80×10^{-3}	
	3-OMs	9.70×10^{-5}	9.12×10^{-5}
		8.63×10^{-5}	
		9.04×10^{-5}	
80E20W	4-OTs	3.97×10^{-2}	4.07×10^{-2}
		3.97×10^{-2}	
		4.09×10^{-2}	
		4.08×10^{-2}	
	3-OTs	1.96×10^{-3}	1.94×10^{-3}
		1.99×10^{-3}	
		1.87×10^{-3}	
	1-OTs	7.84×10^{-5}	8.15×10^{-5}
		8.07×10^{-5}	
		7.88×10^{-5}	
		8.47×10^{-5}	
		8.49×10^{-5}	

1. Kinetics of the Solvolyses of Fluoro-Substituted Benzhydryl Derivatives

Table 1.6. (continued)

solvent[a]	benzhydrylium derivative	k_1 (individual)/ s^{-1}	k_1 (average)/s^{-1}
80E20W	4-OMs	3.49×10^{-2}	3.37×10^{-2}
		3.33×10^{-2}	
		3.36×10^{-2}	
		3.29×10^{-2}	
	3-OMs	1.39×10^{-3}	1.35×10^{-3}
		1.38×10^{-3}	
		1.33×10^{-3}	
		1.31×10^{-3}	
	1-OMs	5.24×10^{-5}	5.27×10^{-5}
		5.31×10^{-5}	
80E20W	7-Br	9.84×10^{-4}	9.47×10^{-4}
		9.45×10^{-4}	
		9.16×10^{-4}	
		9.44×10^{-4}	
	4-Br	2.29×10^{-5}	2.30×10^{-5}
		2.32×10^{-5}	
	5-Br	3.98×10^{-5}	3.98×10^{-5}
		3.98×10^{-5}	
100M	7-OTs	7.40×10^{-1}	8.33×10^{-1}
		7.78×10^{-1}	
		9.81×10^{-1}	
	4-OTs	2.11×10^{-2}	2.07×10^{-2}
		2.01×10^{-2}	
		2.09×10^{-2}	
		2.07×10^{-2}	

1. Kinetics of the Solvolyses of Fluoro-Substituted Benzhydryl Derivatives

Table 1.6. (continued)

solvent[a]	benzhydrylium derivative	k_1 (individual)/ s^{-1}	k_1 (average)/s^{-1}
100M	3-OTs	1.12×10^{-3}	1.13×10^{-3}
		1.09×10^{-3}	
		1.18×10^{-3}	
	1-OTs	5.42×10^{-5}	5.51×10^{-5}
		5.59×10^{-5}	
100M	7-Br	5.76×10^{-4}	5.75×10^{-4}
		5.75×10^{-4}	
		5.77×10^{-4}	
		5.73×10^{-4}	
80M20W	4-Br	1.89×10^{-4}	1.90×10^{-4}
		1.92×10^{-4}	
100TFE	3-OTs	7.07×10^{-2}	7.99×10^{-2}
		8.40×10^{-2}	
		8.90×10^{-2}	
		7.61×10^{-2}	
	1-OTs	1.78×10^{-3}	1.73×10^{-3}
		1.69×10^{-3}	
		1.76×10^{-3}	
		1.69×10^{-3}	
100TFE	3-OMs	9.58×10^{-2}	9.21×10^{-2}
		8.28×10^{-2}	
		9.12×10^{-2}	
		9.85×10^{-2}	
	1-OMs	1.69×10^{-3}	1.78×10^{-3}
		1.91×10^{-3}	
		1.76×10^{-3}	

1. Kinetics of the Solvolyses of Fluoro-Substituted Benzhydryl Derivatives

Table 1.6. (continued)

solvent[a]	benzhydrylium derivative	k_1 (individual)/ s^{-1}	k_1 (average)/s^{-1}
100TFE	7-Br	7.55×10^{-2}	7.27×10^{-2}
		7.77×10^{-2}	
		6.49×10^{-2}	
	4-Br	1.52×10^{-3}	1.49×10^{-3}
		1.53×10^{-3}	
		1.48×10^{-3}	
		1.43×10^{-3}	
	5-Br	2.58×10^{-3}	2.36×10^{-3}
		2.20×10^{-3}	
		2.31×10^{-3}	
	3-Br	2.66×10^{-5}	2.54×10^{-5}
		2.41×10^{-5}	
100TFE	7-Cl	2.17×10^{-2}	2.10×10^{-2}
		2.18×10^{-2}	
		2.02×10^{-2}	
		2.01×10^{-2}	
	4-Cl	3.76×10^{-4}	3.87×10^{-4}
		3.81×10^{-4}	
		4.05×10^{-4}	

[a] Mixtures of solvents are given as (v/v); solvents: A = acetone, AN = acetonitrile, E = ethanol, M = methanol, TFE = 2,2,2-trifluoroethanol, W = water. [b] Stopped flow kinetics.

The Eyring and Arrhenius parameters were determined by measuring the solvolysis rate constants k_s of **7**-Br and **3**-OTs in 80E20W at different temperatures. Plots of ln k_s vs. 1/T (T in K) yielded the activation energy E_a as slope/R and lg A as intercept × lg e. Plots of ln (k_s/T) vs. 1/T (T in K) yielded the activation enthalpy $\Delta H^\ddagger$ as −slope/R and $\Delta S^\ddagger$ as (intercept−ln(k_B/h))/R.

1. Kinetics of the Solvolyses of Fluoro-Substituted Benzhydryl Derivatives

Table 1.7. Rate constants k_s for the solvolysis of **1-Br** and **3-OTs** in 80E20W at different temperatures.

$T/°C$	k_s/s^{-1} (7-Br)	$T/°C$	k_s/s^{-1} (3-OTs)
3.0	6.10×10^{-5}	2.1	1.29×10^{-4}
15.0	2.85×10^{-4}	15.0	6.43×10^{-4}
20.0	5.45×10^{-4}	20.0	1.16×10^{-3}
25.0	9.47×10^{-4}	25.0	1.94×10^{-3}
28.7	1.56×10^{-3}	30.0	3.67×10^{-3}
38.3	4.33×10^{-3}	38.2	8.32×10^{-3}

Table 1.8. Activation parameters for the solvolyses of **1-Br** and **3-OTs** in 80E20W.

	7-Br	3-OTs
$\Delta H^{\ddagger}/\text{kJ mol}^{-1}$	84.2 ± 0.7	79.8 ± 0.8
$\Delta S^{\ddagger}/\text{J mol}^{-1}\text{ K}^{-1}$	-20.2 ± 2.4	-28.9 ± 2.7
$E_a/\text{kJ mol}^{-1}$	86.6 ± 2.4	82.2 ± 0.8
$\lg A$	12.2 ± 0.1	11.7 ± 0.1

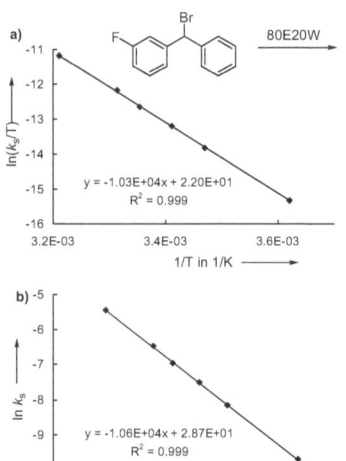

Figure 1.3. Eyring (a) and Arrhenius (b) plot for **7-Br** in 80E20W.

1. Kinetics of the Solvolyses of Fluoro-Substituted Benzhydryl Derivatives

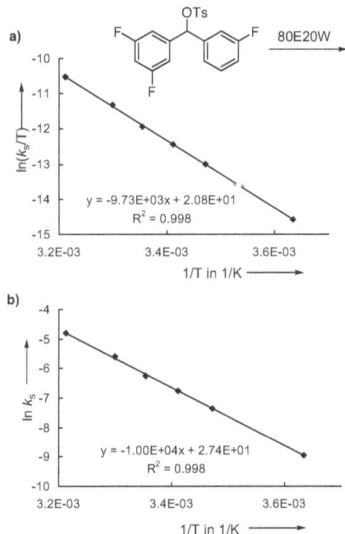

Figure 1.4. Eyring (a) and Arrhenius (b) plot for **3-OTs** in 80E20W.

1. Kinetics of the Solvolyses of Fluoro-Substituted Benzhydryl Derivatives

1.7. References

(1) For reviews on carbocations, see: a) Streitwieser Jr., A. *Solvolytic Displacement Reactions*, McGraw-Hill, New York, **1962**. b) *Carbonium Ions* (Eds.: Olah, G. A.; Schleyer, P. v. R.), Interscience, New York, **1968-1976**; Vols. 1-5. c) Vogel, P. *Carbocation Chemistry*, Elsevier, Amsterdam, 1985. d) *Advances in Carbocation Chemistry* (Ed.: Creary, X.), JAI Press, Greenwich (CT), **1989**, Vol. 1. e) *Advances in Carbocation Chemistry* (Ed.: Coxon, J. M.), JAI Press, Greenwich (CT), **1995**, Vol. 2. f) Katritzky, A. R. *Chem. Soc. Rev.* **1990**, *19*, 83-105. g) McClelland, R. A. in *Reactive Intermediate Chemistry* (Eds.: Moss, R. A.; Platz, M. S.; Jones Jr., M.), Wiley-Interscience: Hoboken (NJ), **2004**, Chapter 1.

(2) a) http://www.cup.lmu.de/oc/mayr/DBintro.html. b) Mayr, H.; Bug, T.; Gotta, M. F.; Hering, N.; Irrgang, B.; Janker, B.; Kempf, B.; Loos, R.; Ofial, A. R.; Remennikov, G.; Schimmel, H. *J. Am. Chem. Soc.* **2001**, *123*, 9500-9512. c) Mayr, H.; Ofial, A. R. *Pure Appl. Chem.* **2005**, *77*, 1807-1821. d) Mayr, H.; Ofial, A. R. *J. Phys. Org. Chem.* **2008**, *21*, 584-595. e) Mayr, H.; Patz, M. *Angew. Chem. Int. Ed.* **1994**, *33*, 938-957. f) Mayr, H.; Kempf, B.; Ofial, A. R. *Acc. Chem. Res.* **2003**, *36*, 66-77.

(3) a) Denegri, B.; Streiter, A.; Juric, S.; Ofial, A. R.; Kronja, O.; Mayr, H. *Chem. Eur. J.* **2006**, *12*, 1648-1656. b) Denegri, B.; Ofial, A. R.; Juric, S.; Streiter, A.; Kronja, O.; Mayr, H. *Chem. Eur. J.* **2006**, *12*, 1657-1666. c) Denegri, B.; Kronja, O. *J. Org. Chem.* **2007**, *72*, 8427-8433. d) Denegri, B.; Kronja, O. *J. Phys. Org. Chem.* **2009**, *22*, 495-503. e) Denegri, B.; Kronja, O. *J. Org. Chem.* **2009**, *74*, 5927-5933. e) Streidl, N.; Denegri, B.; Kronja, O.; Mayr, H. *Acc. Chem. Res.* **2010**, *43*, 1537-1549.

(4) Sailer, C.; Ammer, J.; Nolte, C.; Pugliesi, I.; Mayr, H.; Riedle, E. lecture presented at the *XXIV Int. Conf. on Photochemistry*, Toledo (E), 2009.

(5) For a review on the preparation and application of sulfonates see: Stang P. J.; Hanack M.; Subramanian L. R. *Synthesis* **1982**, 85-102.

(6) Cheeseman G. W. H.; Poller, R. C. *J. Chem. Soc.* **1962**, 5277-5280.

(7) Bentley, W. T.; Christl, M.; Norman, S. J. *J. Org. Chem.* **1991**, *56*, 6240-6243.

1. Kinetics of the Solvolyses of Fluoro-Substituted Benzhydryl Derivatives

(8) For reviews on common-ion rate depression, see: a) Raber, D. J.; Harris, J. M.; Schleyer P. v. R. in *Ions and Ion Pairs in Organic Reactions* (Ed.: Szwarc M.), Wiley, New York, **1974**, Vol. 2. b) Stang, P. J.; Rappoport, Z.; Hanack, M.; Subramanian, L. R. *Vinyl Cations*, Academic Press, New York, **1979**, Chapter 6, pp 337-338. c) Rappoport Z. in *Reactive Intermediates* (Ed.: Abramovitch R. A.), Plenum Press: New York, **1983**, Vol. 3, pp 583-594. d) Kitamura, T.; Taniguchi, H.; Tsuno Y. in *Dicoordinated Carbocations* (Eds.: Rappoport, Z.; Stang, P. J.), Wiley, Chichester, UK, **1997**, pp 321-376. e) Streidl, N.; Antipova, A.; Mayr, H. *J. Org. Chem.* **2009**, *74*, 7328-7334.

(9) a) Cowie, G. R.; Fitches, H. J. M.; Kohnstam, G. *J. Chem. Soc.* **1963**, 1585-1593. b) Fox, J. R.; Kohnstam, G. *J. Chem. Soc.* **1963**, 1593-1598.

(10) a) Fox, J. R.; Kohnstam, G. *Proc. Chem. Soc.* **1964**, 115-116. b) Nishida, S. *J. Org. Chem.* **1967**, *32*, 2692-2695. c) Nishida, S. *J. Org. Chem.* **1967**, *32*, 2695-2697. d) Nishida, S. *J. Org. Chem.* **1967**, *32*, 2697-2701. e) Tsuno, Y.; Fujio, M. *Adv. Phys. Org. Chem.* **1999**, *32*, 267-385.

(11) Hansch, C.; Leo, A.; Taft, R. W. *Chem. Rev.* **1991**, *91*, 165-195.

(12) Schade, C.; Mayr, H. *Tetrahedron* **1988**, *44*, 5761-5770.

(13) Kokura, A.; Tanaka, S.; Ikeno, T.; Yamada, T. *Org. Lett.* **2006**, *8*, 3025-3027.

2. Nucleofugality and Nucleophilicity of Fluoride in Protic Solvents

2.1. Introduction

Fluorine has so far only been found in 30 natural products (status 2004),[1-3] while the other halides have been detected in several thousand molecules synthesized by nature. On the other hand, fluoro-substituted compounds have become highly important in medicinal and agricultural chemistry, [4-6] and 20-25 % of the drugs in the pharmaceutical pipeline contain at least one fluorine atom.[5] Since the van der Waals radius of fluorine (1.47 Å) is between that of oxygen (1.52 Å) and hydrogen (1.20 Å), incorporation of fluorine into a biologically active substances strongly affects the electronic properties without large changes of its structure. Bioavailability, lipophilicity, blood brain barrier permeability, and metabolic stability of a pharmaceutically active molecule can, therefore, be customized by incorporation of fluorine. However, incorporation of fluorine into complex organic molecules is often a challenging task for the synthetic chemist.[7-10] Nucleophilic substitutions with [^{18}F]-fluoride are the key steps in various syntheses of radiopharmaceuticals used in positron emission tomography (PET).[11-15] Though the low reactivity of the C-F bond has already been recognized by Ingold[16] and Hughes[17], only few quantitative data on the leaving group abilities[18-24] and the nucleophilic reactivities of fluoride are available.[25,26] In recent years, the Kronja group and the Mayr group have introduced a novel approach to analyze leaving group abilities in solvolysis reactions.[27-33] By using benzhydrylium ions (Introduction Table 1, Tabel 2.1) of variable stabilization as reference electrofuges, it became possible to compare nucleofugalities of anions and neutral leaving groups in different solvents over a wide range of reactivity. For the correlation of the solvolysis rate constants k_1 (s^{-1}) equation 2.1 was employed, where carbocations are characterized by the electrofugality parameter E_f, and combinations of leaving groups and solvents are characterized by the nucleofuge-specific parameters N_f and s_f.

$$\lg k_1 \ (25\ °C) = s_f(N_f + E_f) \qquad (2.1)$$

s_f, N_f : nucleofuge-specific parameters

E_f : electrofuge-specific parameter.

2. Nucleofugality and Nucleophilicity of Fluoride in Protic Solvents

So far fluoride was not incorporated into these scales which presently include parameters for more than 100 leaving group/solvent pairs. In this thesis, an efficient synthesis of a series of substituted benzhydryl fluorides **11-15**-F, the rate constants (k_1) for their solvolyses in various solvents, and the rates (k_{-1}) of the reactions of fluoride ions with benzhydrylium ions (Table 2.1, Scheme 2.1) in a variety of solvents are reported.

Scheme 2.1. Ionization of benzhydryl fluorides

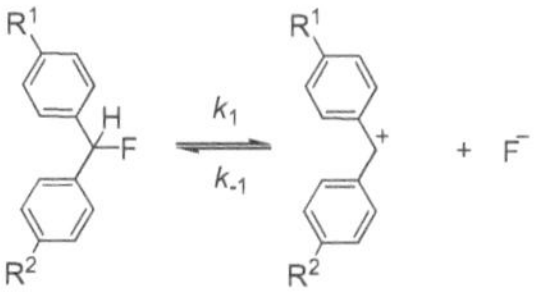

Table 2.1. Benzhydrylium ions employed in this study.

Electrophile/Electrofuge		R^1	R^2	E_f	E
9^+	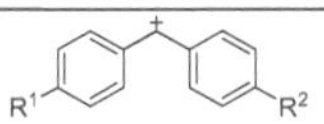	H	H	−6.03	5.90
11^+		Me	Me	−3.44	3.63
12^+		OMe	H	−2.09	2.11
13^+		OMe	Me	−1.23	1.48
14^+		OMe	OPh	−0.86	0.61
15^+		OMe	OMe	0.00	0.00
16^+	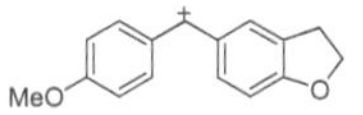			0.61	−0.83
17^+	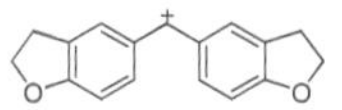			1.07	−1.36
18^+	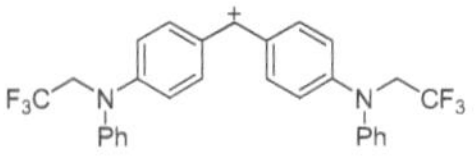			1.79	−3.14
19^+	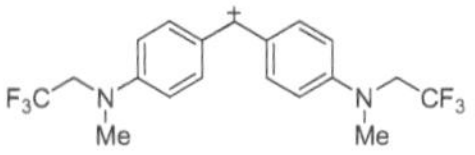			3.13	−3.85
20^+	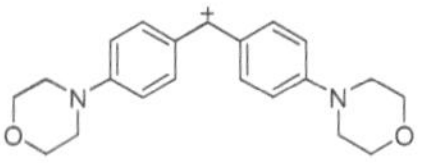			3.03	−5.53
21^+		NMe$_2$	NMe$_2$	4.84	−7.02

2. Nucleofugality and Nucleophilicity of Fluoride in Protic Solvents

2.2. Results

Synthesis of Benzhydryl Fluorides

While Swain et al. used anhydrous hydrofluoric acid for the preparation of trityl fluoride and the parent benzhydryl fluoride (9-F) from the corresponding alcohols,[24] Ando avoided the use of hydrofluoric acid by treating benzhydryl bromide with silver fluoride dispersed on calcium fluoride (Scheme 2.2).[34] Following this procedure, the symmetrical benzhydryl fluorides **11-F** and **15-F** were synthesized, isolated, and characterized. The liquid **11-F** was purified by distillation under high vacuum, and **15-F** was recrystallized from dichloromethane/diethyl ether. The other benzhydryl fluorides were not isolated because they decomposed during distillation and did not crystallize readily. They were synthesized in acetonitrile solution (0.2 M) and used for kinetic investigations without prior evaporation of the solvent. In general, great care has to be taken to exclude traces of water and acid during synthesis and handling of these compounds, since traces of acid lead to autocatalytic decomposition of the benzhydryl fluorides.[20]

Scheme 2.2. Synthesis of benzhydryl fluorides **11-15-F**. The substitution patterns defined by **11-15** are analogous to those shown for the benzhydrylium ions in Table 2.1.

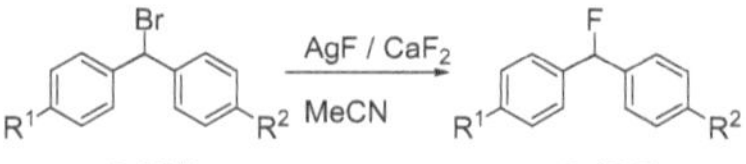

$$\text{9-15-Br} \xrightarrow{\text{AgF / CaF}_2}_{\text{MeCN}} \text{9-15-F}$$

Kinetics of Benzhydryl Fluoride Solvolyses

When compounds **11-15-F** were dissolved in aqueous or alcoholic media, an increase of conductance was observed. As the solvolyses (Scheme 2.3) of alkyl fluorides are prone to autocatalysis,[20] amines were added to trap the protons and to deprotonate the released hydrofluoric acid quantitatively, thus ensuring a linear dependence of the conductance on the reaction progress. Calibration experiments, i.e., stepwise addition of the rapidly solvolyzing benzhydryl fluoride **14-F** to a solution of 80E20W containing 0.08 M piperidine, showed a linear correlation between the initial concentration of the benzhydryl fluoride **14-F** and the conductance at the end of the reaction within the investigated concentration range (see experimental section). As a consequence, monoexponential

2. Nucleofugality and Nucleophilicity of Fluoride in Protic Solvents

increases of the conductance (G) were observed during the solvolysis reaction and the first-order rate constants k_1 (Table 2.2) were obtained by fitting the time dependent conductance G to the monoexponential function (eq. 2.2).

$$G = G_\infty(1 - e^{-k_1 t}) + C \qquad (2.2)$$

The majority of ionization rate constants (k_1) were determined at least at two different concentrations of piperidine (0.08 M to 0.16 M, see Experimental Section); for one system (solvolysis of **14-F** in 60AN40W) 2,6-lutidine was additionally used as additive for the determination of k_1. In all cases the rate constants varied only within the typical experimental error margin (0-7 %).

Scheme 2.3. Simplified solvolysis scheme for S_N1 reactions.

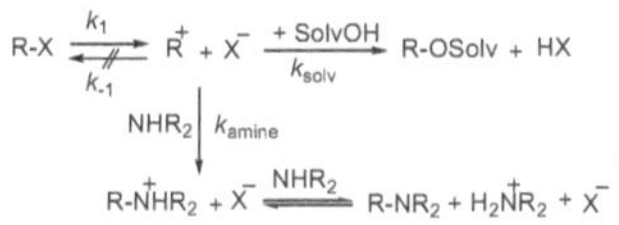

$$\text{R-X} \underset{k_{-1}}{\overset{k_1}{\rightleftharpoons}} \text{R}^+ + \text{X}^- \xrightarrow[k_{\text{solv}}]{+ \text{SolvOH}} \text{R-OSolv} + \text{HX}$$

$$\text{NHR}_2 \Bigg| k_{\text{amine}}$$

$$\text{R-}\overset{+}{\text{N}}\text{HR}_2 + \text{X}^- \xrightleftharpoons[]{\text{NHR}_2} \text{R-NR}_2 + \text{H}_2\overset{+}{\text{N}}\text{R}_2 + \text{X}^-$$

The observation of first-order kinetics, and the fact that the observed first-order rate constants are independent of the nature and concentration of the added amines (Figure 2.1) indicates that the observed first-order rate constants equal k_1. If common-ion return would occur (k_{-1} [X$^-$] ≈ k_{solv}), an increase of the piperidine concentration would lead to an increase of the overall rate because trapping of the carbocation by the amine (k_{amine}) would suppress the ion recombination. The second-order rate law for an S_N2 reaction requires a linear increase of k_{obs} with the concentration of the amine, which can be excluded from the data shown in Figure 2.1.

2. Nucleofugality and Nucleophilicity of Fluoride in Protic Solvents

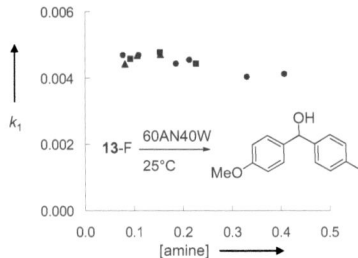

Figure 2.1. Observed rate constants k_1/s^{-1} for the solvolysis of **13-F** in 60AN40 at various concentrations of amine (▲ for triethylamine, ■ for 2,6-lutidine, ● for piperidine).

Table 2.2. Solvolysis rate constants (25 °C) of the benzhydryl fluorides **11-15-F** in different solvents.

solvent[a]	Ar$_2$CHF	k_1/s^{-1}
90A10W	15-F	9.26×10^{-5}
80A20W	15-F	1.20×10^{-3}
	14-F	1.43×10^{-4}
80AN20W	15-F	7.90×10^{-3}
	14-F	1.11×10^{-3}
	13-F	4.85×10^{-4}
100E	15-F[b]	4.26×10^{-3}
	14-F	5.63×10^{-4}
	13-F	1.63×10^{-4}
60A40W	14-F	3.55×10^{-3}
	13-F	1.87×10^{-3}
	12-F	3.80×10^{-4}
	11-F	3.28×10^{-5}
100M	15-F	3.92×10^{-2}
	14-F	5.49×10^{-3}
	13-F	1.99×10^{-3}
	12-F	3.43×10^{-4}

2. Nucleofugality and Nucleophilicity of Fluoride in Protic Solvents

Table 2.2. (continued)

solvent [a]	Ar$_2$CHF	k_1/s^{-1}
80E20W	15-F	7.77 × 10^{-2}
	14-F	1.42 × 10^{-2}
	13-F	4.34 × 10^{-3}
	12-F	7.82 × 10^{-4}
	11-F	5.75 × 10^{-5}
	9-F [c]	2.75 × 10^{-7}
60AN40W	15-F	8.35 × 10^{-2}
	14-F	9.28 × 10^{-3}
	13-F [d]	4.66 × 10^{-3}
	12-F	1.02 × 10^{-3}
	11-F	9.99 × 10^{-5}

[a] Mixtures of solvents are given as (v/v); solvents: A = acetone, AN = acetonitrile, E = ethanol, M = methanol, W = water. [b] Eyring activation parameters: $\Delta H^‡$ = 62.5 kJ mol^{-1}, $\Delta S^‡$ = −81.0 J mol^{-1} K^{-1}. [c] rate constant was determined by Swain et al.[24] [d] Eyring activation parameters: $\Delta H^‡$ = 64.1 kJ mol^{-1}, $\Delta S^‡$ = −74.7 J mol^{-1} K^{-1}

Plots of lg k_1 for the solvolyses of **11-15-F** in various solvents vs. the electrofugality parameters E_f of the benzhydrylium ions are linear (Figure 2.2), indicating the applicability of equation 2.1. From these correlations, one can extract the nucleofugality parameters, N_f as the negative intercepts on the abscissa (E_f axis) and the s_f parameters as the slopes of the correlation lines (Table 2.3).

2. Nucleofugality and Nucleophilicity of Fluoride in Protic Solvents

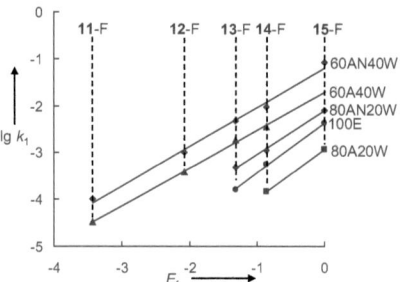

Figure 2.2. Plots of lg k_1 for the solvolysis reactions of various benzhydryl fluorides vs. the electrofugalities E_f. The correlation lines for 80E20W and 100M are shown in the Experimental Section. Mixtures of solvents are given as (v/v); solvents: A = acetone, AN = acetonitrile, E = ethanol, M=methanol, W = water.

The calculated rate constant $k_{1,\text{calcd}} = 2.23 \times 10^{-7}\ \text{s}^{-1}$ for the solvolysis of **9-F** calculated from equation 2.1 using the nucleofugality parameters for fluoride in 80E20W (Table 2.3) and the electrofugality parameter for the unsubstituted benzhydryl cation ($E_f = -6.03$)[31] is in excellent agreement with the previously reported experimental rate constant of $2.75 \times 10^{-7}\ \text{s}^{-1}$,[24] which demonstrates the power of equation 2.1 and the practicability of these parameters.

2. Nucleofugality and Nucleophilicity of Fluoride in Protic Solvents

Table 2.3. Nucleofugality parameters N_f and s_f for the fluoride in various solvents.

	N_f	s_f
80A20W	−2.72	1.07
80AN20W	−2.28	0.93
100E	−2.21	1.07
60A40W	2.14	0.81
60AN40W	−1.44	0.84
100M	−1.43	0.99
80E20W	−1.20	0.92

Nucleophilicity of Fluoride Anions in Various Solvents

It is well known that nucleophilicity is not simply the reverse of nucleofugality[31] and, therefore, the nucleophilic reactivity of the fluoride anion has been determined separately. Previously, the nucleophilicity parameters N and s_N of chloride and bromide in various solvents were determined[35] by measuring the rate constants k_{-1} (M^{-1} s^{-1}) of their reactions with benzhydrylium ions and correlation of the data by equation 2.3.

$$\lg k_{-1}\,(20\ °C) = s_N(N + E) \qquad (2.3)$$

s_N, N: nucleophile-specific parameters
E: electrophile-specific parameter.

Now laser-flash photolysis and stopped-flow techniques are used to characterize the nucleophilic reactivity of fluoride ions by measuring the rates of its reactions with benzhydryl cations 11^+-21^+ in a series of solvents.

The benzhydryltriphenylphosphonium tetrafluoroborates **11-17-PPh$_3$**, which were employed as precursors for the laser-flash photolytic generation of benzhydryl cations, were prepared by treating corresponding benzhydrols with equimolar amounts of triphenylphosphane and aqueous tetrafluoroboric acid, followed by heating to 140-180 °C.[36] The benzhydryltributylphosphonium tetrafluoroborates **18-PBu$_3$** and **19-PBu$_3$** were prepared in acetonitrile solution by adding tributylphosphane to the corresponding benzhydrylium tetrafluoroborates **18$^+$**, **19$^+$** until complete decolorization was achieved. The

2. Nucleofugality and Nucleophilicity of Fluoride in Protic Solvents

stabilized benzhydrylium tetrafluoroborates **18⁺-21⁺** were prepared as previously described.[37]

Scheme 2.4. Precursors for the laser-flash photolytic generation of **11⁺-19⁺**. The substituents X and Y are defined in Table 2.1.

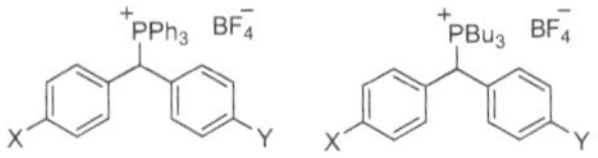

N°-PPh₃   N°-PBu₃

The choice of suitable sources of fluoride anions is not trivial. Unlike tetrabutylammonium chloride and bromide, which was used in the previous study,[35] anhydrous tetrabutylammonium fluoride is prone to elimination and is only stable for several hours at room temperature.[38] Therefore, the commercially available tetrabutylammonium fluoride trihydrate was used, which did not allow us to characterize fluoride in anhydrous solvents. For the kinetic investigations in methanol, water, and 10AN90W potassium and cesium fluoride were used, which were sufficiently soluble in these solvents. Methanol was the only anhydrous solvent, where both potassium fluoride and cesium fluoride could be used as fluoride source. In other anhydrous solvents, such as acetonitrile, the solubility of alkali fluorides is too low for kinetic studies. Tetrabutylammonium fluoride trihydrate was exclusively used as fluoride source for the investigations in the aqueous solvent mixtures (98AN2W, 90AN20W, 80AN20W, 60AN40W, 80E20W) because a phase separation was observed when trying to dissolve potassium fluoride in these mixtures.

The benzhydrylium ions **11⁺-19⁺** were generated in these solvents by irradiation of **11-17**-PPh₃ or **18,19**-PBu₃ with a 7 ns laser pulse from the fourth harmonic of a Nd/YAG Laser (266 nm).[36,39] In the absence of added nucleophiles, typically monoexponential decays of the absorbances of **11⁺-19⁺** were observed. Fitting to the monoexponential equation $A_t = A_0 e^{-k_{obs}t} + C$ provided the first-order rate constants for the reactions of **11⁺-19⁺** with the solvent (k_{solv}), which are listed in Table 2.4. When the carbocations were generated in the presence of added fluoride, the observed rate constants increased linearly with the concentration of fluoride (Figure 2.3). As expressed by equation 2.4, the observed pseudo-first-order rate constants k_{obs} are the sum of a second-order term for the reactions of the carbocations with halide ions (k_{-1} in Scheme 2.3) and a first-order term for the reactions of the carbocations with the solvents (k_{solv}).

2. Nucleofugality and Nucleophilicity of Fluoride in Protic Solvents

$$k_{obs} = k_{-1} [F^-] + k_{solv} \qquad (2.4)$$

Plots of k_{obs} vs. the fluoride concentrations resulted in linear correlations according to equation 2.4 as exemplified in Figure 2.3. The second-order rate constants k_{-1} for the reactions with fluoride listed in Table 2.4 were obtained from the slopes of these plots. The intercepts correspond to the background reactions with the solvent (k_{solv}) which were also determined independently (Table 2.4). Table 2.19 in the Experimental Section demonstrates the good agreement of the experimental data with the values calculated from equation 2.3 using the previously published N_1 and s_N parameters of the solvents.[40]

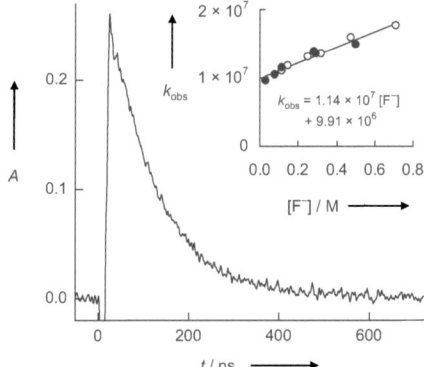

Figure 2.3. Decay of the absorption of (**15**$^+$) in methanol, observed at 500 nm in the presence of 0.3 M CsF. Plot of k_{obs} against [Nu] = [KF] (○) or [CsF] (●) in the inset.

2. Nucleofugality and Nucleophilicity of Fluoride in Protic Solvents

Table 2.4. Rate constants for the reactions of benzhydrylium ions **11⁺-21⁺** with fluoride ($k_{-1}/M^{-1}s^{-1}$) and pure solvent (k_{solv}/s^{-1}). Mixtures of solvents are given as (v/v); solvents: AN = acetonitrile, M = methanol, W = water.

Ar$_2$CH⁺	Solvent	Rxn. with F⁻ $k_{-1}/M^{-1} s^{-1}$	Rxn. with solvent k_{solv}/s^{-1}
17⁺	90AN10W	2.89 × 10⁶	7.05 × 10³
16⁺		6.03 × 10⁶	2.12 × 10⁴
15⁺		1.85 × 10⁷	6.78 × 10⁴
17⁺	100M	1.56 × 10⁶	7.39 × 10⁵
16⁺		4.12 × 10⁶	2.44 × 10⁶
15⁺		1.14 × 10⁷	9.91 × 10⁶ ᵃ
14⁺		ᵇ	1.97 × 10⁷
21⁺	98AN2W	1.97 × 10³	ᶜ
20⁺		2.95 × 10⁴	ᶜ
19⁺		4.17 × 10⁵	ᶜ
18⁺		2.31 × 10⁶	ᶜ
17⁺		1.12 × 10⁸	ᶜ
16⁺		1.40 × 10⁸	ᶜ
15⁺		2.69 × 10⁸	ᶜ
12⁺		2.40 × 10⁹	3.85 × 10⁵
11⁺		8.63 × 10⁹	6.18 × 10⁶
17⁺	80AN20W	6.72 × 10⁵	9.38 × 10³
16⁺		1.72 × 10⁶	3.08 × 10⁴
15⁺		5.10 × 10⁶	9.49 × 10⁴
14⁺		9.76 × 10⁶	2.52 × 10⁵
17⁺	60AN40W	1.66 × 10⁵	1.14 × 10⁴
16⁺		3.94 × 10⁵	3.16 × 10⁴
15⁺		1.82 × 10⁶	9.61 × 10⁴
14⁺		4.36 × 10⁶	2.81 × 10⁵
13⁺		9.04 × 10⁶	1.03 × 10⁶
11⁺		2.0 × 10⁸	3.35 × 10⁷

2. Nucleofugality and Nucleophilicity of Fluoride in Protic Solvents

Table 2.4. (continued)

Ar_2CH^+	Solvent	Rxn. with F^- $k_{-1}/M^{-1} s^{-1}$	Rxn. with solvent k_{solv}/s^{-1}
15⁺	10AN90W	1.37×10^5	$1.40 \times 10^{5\,d}$
14⁺		3.28×10^5	3.15×10^5
13⁺		1.20×10^6	$1.04 \times 10^{6\,d}$
15⁺	100W	1.09×10^5	$1.79 \times 10^{5\,e}$
13⁺		1.02×10^6	$1.02 \times 10^{6\,e}$
16⁺	80E20W	3.83×10^6	4.87×10^5
15⁺		1.06×10^7	1.66×10^6

[a] Previously $k_{solv} = 8.4 \times 10^6\,s^{-1}$ has been reported Ref. [41]. [b] The rate constant k_{-1} could not be determined because the slope in the plot of [F⁻] versus k_{obs} was too small. [c] k_{solv} was not determined as the expected rate constant is below the measuring range of the laser-flash instrument and determination from the intercept of the plots k_{obs} vs. [F⁻] is too imprecise as $k_{-1}/k_{solv} > 1000$. [d] Previously $k_{solv} = 9.55 \times 10^4\,s^{-1}$ for (15⁺) and $k_{solv} = 7.99 \times 10^5\,s^{-1}$ for (13⁺) in 10AN90W have been reported in Ref. [40]. [e] Previously $k_{solv} = 1.0 \times 10^5\,s^{-1}$ for (15⁺) and $k_{solv} = 7.8 \times 10^5\,s^{-1}$ for (13⁺) in 100W have been reported in Ref. [42].

As the fluoride anion is a weak base, it had to be ensured that fluoride and not hydroxide was the acting nucleophile in aqueous solvents. For the reaction of the bisanisyl carbenium ion (15⁺) with fluoride in water, fluoride concentrations ranging from 0.06 M to 0.91 M were used. From pK_b for fluoride in water (10.9), one calculates concentrations of hydroxide from 0.87×10^{-6} M to 3.4×10^{-6} M. With equation 2.3 and the published nucleophilicity parameter for hydroxide in water ($N = 10.47$, $s_N = 0.61$)[43] hypothetical pseudo-first-order rate constants $k_{1\Psi}$ (OH⁻) = 2.12 to 8.25 s⁻¹ can be calculated for the reaction of 15⁺ with hydroxide at the above-mentioned concentrations of fluoride. As shown in the Experimental Section, rate constants k_{obs} of $1.79 \times 10^5\,s^{-1}$ to $2.72 \times 10^5\,s^{-1}$ were observed for the reaction of the bisanisyl carbenium ion (15⁺) with fluoride. Therefore, the nucleophilic reactivity of the hydroxide anion can be neglected for the evaluation of the kinetic experiments.

Further evidence that fluoride, and not hydroxide, was the active nucleophile was obtained by a ¹H-NMR spectroscopic product analysis. A solution of 4,4'-dimethylbenzhydryl chloride (11-Cl) in deuterated acetonitrile was combined with a solution of tetrabutylammonium fluoride trihydrate in aqueous acetonitrile to yield a solution of 0.15 M 4,4-dimethylbenzhydryl chloride (11-Cl) and 0.54 M tetrabutylammonium fluoride in 60 %

2. Nucleofugality and Nucleophilicity of Fluoride in Protic Solvents

aqueous acetonitrile (60AN40W). In line with a calculated solvolysis rate constant of $k_1 \approx$ 2.42 s^{-1} at 25 °C,[44] **11**-Cl was not observable in a ^{1}H-NMR spectrum taken immediately after mixing the two solutions. As depicted in Figure 2.4, the characteristic doublet at δ = 6.46 (d, 1 H, $^3J_{HF}$ = 46.0 Hz, CHF) for **11-F** and singlet at δ = 5.69 indicated a product ratio of 2.39/1 (**11-F/11-OH**). As the benzhydryl fluoride **11-F** will solvolyze with a half-life of 1.9 h under these conditions (Table 2.2), the observed 4,4'-dimethylbenzhydrol (**11-OH**) cannot arise from hydrolysis of **11-F**, and the observed product ratio reflects the relative reactivities of the benzhydryl cation **11$^+$** toward F$^-$ and H$_2$O (OH$^-$ negligible as discussed above). This ratio shall be compared with the ratio of the absolute rate constants for the reaction of **11$^+$** with F$^-$ and H$_2$O. The second-order rate constant for the reaction of **11$^+$** with fluoride in 60AN40W has been determined (k_{-1} = 2.0 × 10^8 M^{-1} s^{-1})[45]. The rate constant for the reaction of **11$^+$** with 60 % aqueous acetonitrile is 3.35 × 10^7 s^{-1} (see Table 2.4). According to the ^{1}H-NMR spectrum 70 % of **11**-Cl (c$_0$ = 0.15 M) are converted into the benzhydryl fluoride **11-F**. Thus, the average concentration of fluoride during the reaction is 0.49 M (eq. 2.5).

$$[F^-]_{average} = [F^-]_0 - \frac{[Ar_2CHF]}{2} = 0.54M - \frac{[0.70 \times 0.15M]}{2} = 0.49M \quad (2.5)$$

Multiplication of the second order rate constant (k_{-1}) for the reaction of **11$^+$** with [F$^-$]$_{average}$ yields the pseudo-first-order rate constant for the reaction of **11$^+$** with F$^-$, which is divided by k_{solv} to yield an expected product ratio of 2.96/1 (eq. 2.6), in fair agreement with the product ratio observed by ^{1}H-NMR (2.39).

$$\frac{[11-F]}{[11-OH]} = \frac{k_{-1} \times [F^-]_{average}}{k_{solv}} = \frac{2.0 \times 10^8 \times 0.49}{3.31 \times 10^7} = 2.96 \quad (2.6)$$

A more accurate comparison of the kinetic results with the ^{1}H-NMR product analysis is provided in the Experimental Section.

2. Nucleofugality and Nucleophilicity of Fluoride in Protic Solvents

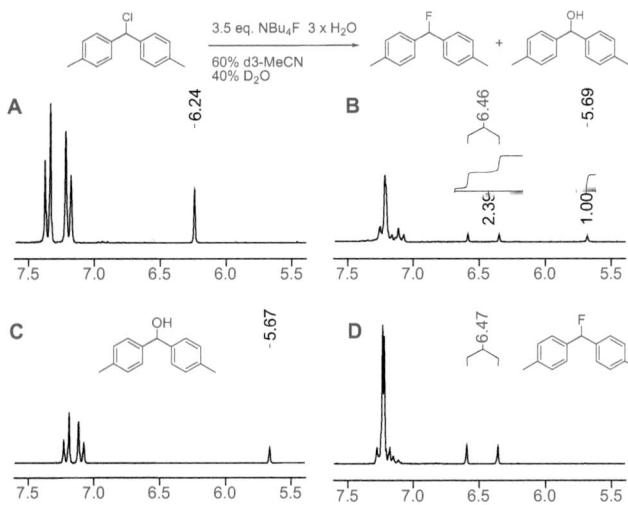

Figure 2.4. ^{1}H-NMR-spectra (200 MHz) recorded for product analysis; **A** 11-Cl in deuterated acetonitrile; **B** spectrum recorded 5 minutes after combining 11-Cl with 3.5 equivalents of tetrabutylammonium fluoride trihydrate in 60AN40W (deuterated solvents); **C** 11-OH in 60AN40W (deuterated solvents); **D** 11-F in deuterated acetonitrile.

When plotting lg k_{-1} for the reactions of F$^-$ with benzhydryl cations against the electrophilicities E of the corresponding benzhydryl cations, linear correlations according to equation 2.3 were obtained (Figure 2.5). In 98AN2W solution, absolute rate constants for the reactions of fluoride with a wide variety of benzhydrylium ions were determined. The rate constants for the reactions of **20$^+$** and **21$^+$** with fluoride determined by the stopped-flow method,[37] lie on the same graph as the rate constants for the reactions of **11$^+$-19$^+$** which were determined by the laser-flash photolysis technique, demonstrating the consistency of the results obtained by the different methods (Figure 2.5). The linear correlation for the rate constants for **17$^+$-21$^+$** bends down as k_{-1} exceeds 10^8 M^{-1} s^{-1} due to the proximity of the diffusion limit. An analogous behavior was observed for numerous other nucleophiles.[46]

2. Nucleofugality and Nucleophilicity of Fluoride in Protic Solvents

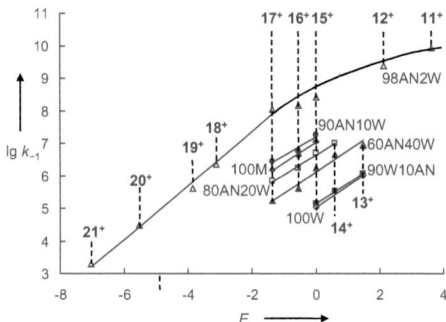

Figure 2.5. Plot of lg k_{-1} for the reactions of benzhydrylium ions with fluoride ions versus their electrophilicity parameters E. Mixtures of solvents are given as (v/v); solvents: AN = acetonitrile, M = methanol, W = water. Two data points for 80E20W were superimposed by data points in methanol (100M).

Only a narrow range of carbocations could be investigated in solvent mixtures containing a higher percentage of water or alcohols. The reactions of fluoride with the more stabilized benzhydrylium ions ($E < -2$) do not proceed quantitatively, and the fast reversible reactions with F⁻ have to compete with the slower but irreversible reactions with water giving rise to kinetics which are difficult to evaluate. On the other hand, the reactions of fluoride with the less stabilized benzhydrylium ions ($E > 2$) are difficult to follow, because for highly electrophilic carbocations the major fraction of k_{obs} (eq. 2.4) is due to the reaction of the benzhydrylium ion with the solvent. As shown for the example of methanol in Figure 2.6 the rate constants for the reactions of benzhydrylium ions **(14-17)⁺** with the solvents are more sensitive toward variation of the electrophiles than those of the reactions of benzhydrylium ions **(14-17)⁺** with fluoride (fluoride has lower s_N parameter in eq. 2.3, see below), with the consequence that the accurate determination of the small contribution of the k_{-1}[F⁻] term becomes difficult.

2. Nucleofugality and Nucleophilicity of Fluoride in Protic Solvents

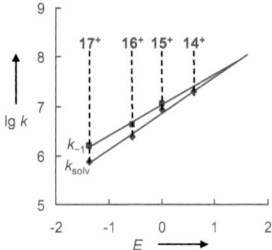

Figure 2.6. Plot of lg k_{-1} (♦) and lg k_{solv} (■) for the reactions of benzhydrylium ions (**14-17**)$^+$ with fluoride ions and methanol versus their electrophilicity parameters E.

Another limitation of the laser-flash photolytic technique is the recombination of the benzhydrylium ions with the phosphine photo-leaving group. For that reason, rate constants $k_{obs} < 10^5$ M^{-1} s^{-1} in 100W and in 10AN90W could not be determined.

Plots of the second-order rate constants k_{-1} for the reactions of benzhydrylium ions (**13-21**)$^+$ with fluoride (Table 2.4) against the electrophilicity parameters E of the benzhydrylium ions were linear (Figure 2.5) and yielded the s_N parameters and the nucleophilicity parameters N of fluoride in methanol and aqueous solvents (Table 2.5). Rate constants $k_{obs} > 1 \times 10^8$ M^{-1} s^{-1} were not used for the correlations as these reactions approach the diffusion limit (see above).

Table 2.5. Nucleophilicity parameters N and s_N for fluoride in various solvents. Mixtures of solvents are given as (v/v); solvents: AN = acetonitrile, M = methanol, W = water.

Solvent	N	s_N
90AN10W	12.27	0.59
80AN20W	11.40	0.59
100M	11.31	0.63
98AN2W	10.88	0.83
60AN40W	9.75	0.63
10AN90W	8.05	0.64
100W	7.68	0.65

2. Nucleofugality and Nucleophilicity of Fluoride in Protic Solvents

2.3. Discussion

With the nucleofugality parameters (N_f, s_f) listed in Table 2.3, it now becomes possible to directly compare the leaving group abilities of fluoride and other common leaving groups in different solvents. As $s_f \approx 1$, a general overview can be derived directly from the N_f parameters, some of which are compared in Table 2.6. Depending on the solvent, fluoride is a slightly weaker or better leaving group than 3,5-dinitrobenzoate ($N_f = -2.05$; $s_f = 1.09$ in 100E).[31]

Table 2.6. Comparison of the nucleofugalities N_f of important leaving groups in different solvents and solvolysis half lives of Ph$_2$CHX in 80E20W at 25 °C.

	DNB [a]	F	Cl	Br
80A20W	−2.34	−2.73	2.03	3.01
EtOH	−2.05	−2.21	1.82	2.93
80E20W	−1.43	−1.20	3.24	4.36
$\tau_{1/2}$ [b]	164 d [c]	29 d	6 min	23 s

[a] 3,5-dinitrobenzoate [b] Solvolysis half life of Ph$_2$CHX (9-X) in 80 % aqueous ethanol (80E20W) at 25 °C [c] calculated by equation 1

Replacing fluoride by chloride accelerates the solvolysis reactions by approximately by 4-5 orders of magnitude, while replacing fluoride by bromide results in an acceleration of approximately 5-6 orders of magnitude. Thus the unsubstituted benzhydryl fluoride will solvolyze in 80E20W with a half-life of a month, whereas the half-life is 6 minutes for benzhydryl chloride, 23 seconds for benzhydryl bromide and only 50 milliseconds for the parent benzhydryl tosylate (not shown in Table 2.6). These findings are in agreement with previous results by Swain and Scott who reported chloride/fluoride ratios of 10^6 and 10^5 for the couples trityl chloride/trityl fluoride (85 % aq. acetone) and *tert*-butyl chloride/*tert*-butyl fluoride (80 % aq. ethanol).[18] The poor nucleofugality of F$^-$ is commonly accounted to the high C-F bond energy. Table 2.7 shows a further reason: The activation entropies of benzhydryl fluorides are considerably more negative than those of benzhydryl chlorides and bromides. Thus, entries 1-3 show that in 60 % aqueous acetonitrile, $\Delta S^{\ddagger}$ becomes more negative in the series Ar$_2$CHBr to Ar$_2$CHCl and Ar$_2$CHF. The same trend, more negative entropy of activation for R-Cl than for R-Br has also been observed in the *tert*-butyl series (entries 4 and 5). The higher degree of solvent orientation needed for the solvation of

2. Nucleofugality and Nucleophilicity of Fluoride in Protic Solvents

fluoride ions is in line with the relative hydration energies for halide ions ($-\Delta H_h°$), which increase from I^- (294 kJ mol^{-1}), Br^- (335 kJ mol^{-1}), Cl^- (366 kJ mol^{-1}) to F^- (502 kJ mol^{-1}).[25]

Table 2.7. Activation parameters for the solvolyses of benzhydryl and *tert*-butyl halides in different solvents.

substrate	solvent[c]	$\Delta H^{\ddagger}$ / kJ mol^{-1}	$\Delta S^{\ddagger}$ / J mol^{-1} K^{-1}
Ar$_2$CHF (13-F)	60AN40W	64.2	−74.5
(3-FC$_6$H$_4$)PhCH-Cl (7-Cl)[a]	60AN40W	83.0	−34.6
(3-FC$_6$H$_4$)PhCH-Br (7-Br)[a]	60AN40W	86.5	0.69
t-butyl chloride[b]	80A20W	90.3	−51.6
t-butyl bromide[b]	80A20W	85.1	−35.0

[a] This work [b] Data from Ref. [47] [c] Mixtures of solvents are given as (v/v); solvents: A = acetone, AN = acetonitrile, E = ethanol, W = water

As illustrated for the dimethyl substituted benzhydryl derivatives tol$_2$CH-X (**11-X**) in Figure 2.7, the leaving group abilities increase significantly in the series $F^- \approx$ DNB $\ll Cl^- < Br^-$, but the exact ranking depends on the solvent.

2. Nucleofugality and Nucleophilicity of Fluoride in Protic Solvents

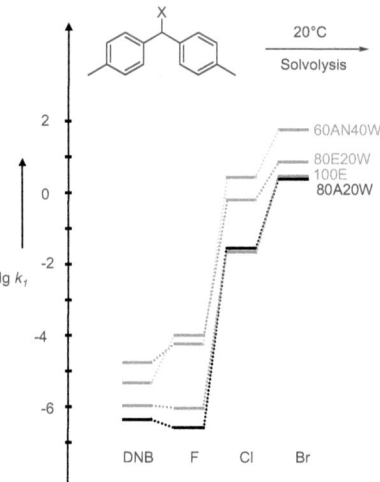

Figure 2.7. Comparison of the solvolysis rates for the reactions of the dimethyl substituted benzhydryl derivative tol$_2$CH-X (**11-X**) with different leaving groups (DNB = 3,5-dinitrobenzoate). Mixtures of solvents are given as (v/v); solvents: AN = acetonitrile, E = ethanol, W = water.

As previously reported for several other leaving groups,[31,48] the sensitivity parameters s_f of fluoride decrease slightly with increasing water content of the solvents. Since the carbocation character is not fully developed in most of the benzhydryl fluoride solvolyses investigated in this work (see below), the trend in s_f might be explained by a smaller degree of charge separation in the transition states in solvents with a high percentage of water. As analogous trends in of s_f are also observed in solvolyses of benzhydryl chlorides and bromides, where the ion combination is diffusion-limited, i.e., where the transition states correspond to the carbocations, other factors must contribute.[31]

Fluoride is not only a poorer nucleofuge than the other halide ions, but also a poorer nucleophile in protic solvents. As shown in Figure 2.8, nucleophilicity increases in the series $F^- < Cl^- < Br^-$ in water, aqueous acetonitrile, and methanol.[49] A strong reduction of the nucleophilic reactivity of fluoride ions by water molecules has also been observed in S_N2 reactions.[25]

2. Nucleofugality and Nucleophilicity of Fluoride in Protic Solvents

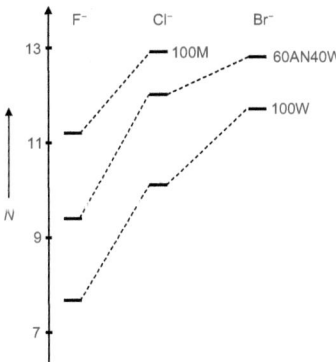

Figure 2.8. Comparison of the nucleophilic reactivities of fluoride with other halide anions in different solvents. Mixtures of solvents are given as (v/v); solvents: AN = acetonitrile, M = methanol, W = water.

The low nucleophilicity of fluoride in protic solvents compared to chloride and bromide accounts for the fact that common-ion return is rarely encountered in S_N1 reactions of alkyl fluorides. As a result, deviations from the first-order rate law, due to reversible ionization towards the end of the kinetic experiments, have not been observed in any of the benzhydryl fluoride solvolyses described above.

A quantitative rationalization for this observation is given in Figure 2.9, where the first-order rate constants of the reactions of various benzhydrylium cations with water in 60 % aqueous acetonitrile are compared with the corresponding pseudo-first-order rate constants with fluoride (i.e., $k_{-1}[F^-]$) at different concentrations of fluoride. One can see that at substrate concentrations which are typical for solvolysis experiments (1.7 mM), the reaction with F^- is approximately 10^2 times slower than the reaction with the solvent. Only when high concentrations of fluoride are employed, the reaction with F^- is faster than the reaction with water as demonstrated for the solvolysis of bis *p*-tolylmethyl chloride [**11-Cl**] in a 0.54 M solution of $nBu_4N^+ F^-$ in 60AN40W (Figure 2.4).

2. Nucleofugality and Nucleophilicity of Fluoride in Protic Solvents

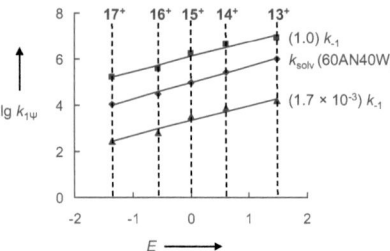

Figure 2.9. Plot of lg $k_{1\psi}$ for the reactions of benzhydrylium ions with aqueous acetonitrile 60AN40W (k_{solv}) and with fluoride ([F⁻] × k_{-1}) at fluoride concentrations of 1.0 M (upper graph) and at [F⁻] = 1.7 mM (lower graph) versus their electrophilicity parameters (E).

2.4. Conclusion

The rate constants for the forward and backward reactions of benzhydryl fluoride ionization can be combined to construct quantitative energy profiles for the solvolysis reactions (Figure 2.10), which differ significantly from those previously derived for benzhydryl chlorides and bromides.

2. Nucleofugality and Nucleophilicity of Fluoride in Protic Solvents

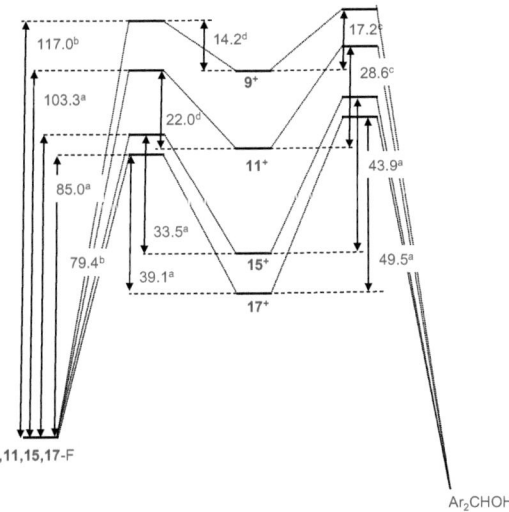

Figure 2.10. Free energy profiles (kJ mol^{-1}) for the solyolyses of differently substituted benzhydrylfluorides in 80AN20W at 25 °C (20 °C for the reactions of 9^+,11^+,15^+ and 17^+ with nucleophiles). ^{a} calculated by equation 2.1. ^{b} direct measurement. ^{c} calculated by equation 2.3.

While S_N1 solvolyses of benzhydryl chlorides and bromides which are commonly investigated at room temperature (i.e., 0.1 s < $\tau_{1/2}$ < 1 d) have carbocation-like transition states (i.e., barrier-free combinations of R^+ with Cl^- or Br^-), solvolysis reactions of benzhydryl fluorides, which proceed with similar rates, typically do not have carbocation-like transition states. As shown in Figure 2.10, the solvolysis of **11-F**, for which a solvolysis half-life of approximately 1 day can be calculated, still yields a carbocation which does not undergo barrier-free recombination with the fluoride ion in aqueous acetonitrile. For the more rapidly ionizing substrates **15-F** ($\tau_{1/2}$ = 88 s) and **17-F** ($\tau_{1/2\ calcd}$ = 9 s) the energy sink for the resulting carbocations is much deeper, and one can extrapolate that similar situations should be encountered for other alkyl halides with comparable ionization rates. One can, therefore conclude that only alkyl fluorides which give much less stabilized carbocations, i.e., substrates RF which require heating to solvolyze within reasonable time periods will ionize via carbocation-like transition states.

2. Nucleofugality and Nucleophilicity of Fluoride in Protic Solvents

2.5. Experimental Section

Silver carbonate

Silver carbonate was prepared according to a literature procedure by combining diluted aqueous solutions of silver nitrate with potassium bicarbonate.[50]

Ando's fluorination agent[34]

The reagent for the fluorination reaction was prepared according to the literature procedure by Ando.[34] Therefore, silver carbonate (10 g, 36 mmol) was mixed with calcium fluoride (40 g, 0.51 mol) and grinded thoroughly. The mixture was transferred to a Teflon round bottom flask equipped with a Teflon stirring bar, and 40 mL of water were added. Then conc. HF (3.0 mL, 73 mmol) was slowly added to the stirred suspension, using a plastic syringe equipped with a Teflon tube. The suspension was stirred for 30 min, followed by evaporation of the water and drying for several hours using high vacuum (1.3 × 10^{-3} mbar) at 40-50 °C. The fluorination agent (49 g, 98 %) was obtained as slightly yellowish free flowing granular powder (Ando reported a colorless powder). The reagent can be stored in an opaque flask under air and can be used for at least a month, storage in an argon filled glove box is recommended. The concentration of active fluorination agent (AgF) is 1.5 mmol/g in the resulting fluorination reagent.

Benzhydryl bromides

The benzhydryl bromides have been used previously but some were only partially characterized.[51-56]

Table 2.8. Benzhydryl bromides employed in this study

Benzhydryl bromide	R^1	R^2	E_f	E
11-Br	Me	Me	−3.44	3.63
12-Br	OMe	H	−2.09	2.11
13-Br	OMe	Me	−1.23	1.48
14-Br	OMe	OPh	−0.86	0.61
15-Br	OMe	OMe	0.00	0.00

2. Nucleofugality and Nucleophilicity of Fluoride in Protic Solvents

Benzhydryl bromide synthesis was performed by refluxing the corresponding benzhydrols with 10 equivalents of acetyl bromide for 15 minutes, followed by evaporation under vacuum. The remaining residue is crystallized in a mixture of dichloromethane and n-pentane (≈1:10). Sometimes good results were also obtained with only 5 equivalents of acetyl bromide. Usually, better results were achieved with the high excess of acetyl bromide, which often yielded crystalline material after evaporation of the acetyl bromide and acetic anhydride.

11-Br was obtained from 4,4'-dimethylbenzhydrol (3.99 g, 18.8 mmol) and acetyl bromide (23.1 g, 188 mmol). The product was isolated as colorless crystals (4.15 g, 80 %). ^{1}H-NMR-spectra are in agreement with literature data. Unlike in Ref [52] the compound could be crystallized.

^{1}H–NMR (300 MHz, CDCl$_3$): δ = 2.35 (s, 6 H, Me), 6.28 (s, 1 H, CHBr), 7.13–7.16 (m, 4 H, ArH), 7.34–7.37 (m, 4 H, ArH).

^{13}C–NMR (75 MHz, CDCl$_3$): δ = 21.2 (Me), 55.9 (CHBr), 128.5 (Ar), 129.3 (Ar), 138.0 (Ar), 138.5 (Ar).

MS (+EI): m/z (%) = 195.2 (100) [M$^+$–Br].

m.p.: 48.0 °C - 48.6 °C

Elemental Analysis: Calculated for C$_{15}$H$_{15}$Br: C, 65.47; H, 5.49.
                    Found: C, 65.70; H, 5.55.

12-Br was obtained from 4-methoxybenzhydrol (2.25 g, 10.5 mmol) and acetyl bromide (12.9 g, 105 mmol). The product was isolated as colorless crystals (2.17 g, 74 %). This compound has been reported previously.[53,54] ^{1}H-NMR-spectra are in agreement with literature data.[57]

^{1}H–NMR (300 MHz, CDCl$_3$): δ = 3.79 (s, 3 H, OMe), 6.30 (s, 1 H, CHBr), 6.84-6.87 (m, 2 H, ArH), 7.25-7.38 (m, 5 H, ArH), 7.45-7.48 (m, 2H, ArH).

^{13}C–NMR (75 MHz, CDCl$_3$): δ = 55.47 (OMe), 55.8 (CHBr), 114.1 (Ar), 128.1 (Ar), 128.5 (Ar), 128.6 (Ar), 129.9 (Ar), 133.5 (Ar), 141.4 (Ar), 159.5 (Ar).

MS (+EI): m/z (%) = 197.2 (100) [M$^+$–Br].

m.p.: 49.5 °C - 50.9 °C

2. Nucleofugality and Nucleophilicity of Fluoride in Protic Solvents

13-Br was obtained from 4-methoxy-4'methylbenzhydrol (2.33 g, 10.2 mmol) and acetyl bromide (13.0 g, 102 mmol). The product was isolated as colorless crystals (2.03 g, 69 %).
^{1}H–NMR (300 MHz, CDCl$_3$): δ = 2.35 (s, 3 H, Me), 3.81 (s, 3 H, OMe), 6.30 (s, 1 H, CHBr), 6.85-6.88 (m, 2 H, ArH), 7.14-7.16 (m, 2 H, ArH), 7.35-7.40 (m, 4 H, ArH).
^{13}C–NMR (75 MHz, CDCl$_3$): δ = 21.2 (Me), 55.5 (OMe), 56.0 (CHBr), 114.0 (Ar), 128.4 (Ar), 129.3 (Ar), 129.8 (Ar), 133.7 (Ar), 138.0 (Ar), 138.5 (Ar), 159.4 (Ar).
MS (+EI): m/z (%) = 211.2 (100) [M$^+$–Br].
m.p.: 39.1 °C - 40.2 °C

14-Br was obtained from 4-methoxy-4'phenoxybenzhydrol (1.98 g, 6.46 mmol) and acetyl bromide (7.95 g, 64.7 mmol). The product was isolated as slightly pink crystals (2.03 g, 67 %).
^{1}H–NMR (300 MHz, CDCl$_3$): δ = 3.81 (s, 3 H, OMe), 6.32 (s, 1 H, CHBr), 6.86-7.11 (m, 7 H, ArH), 7.33-7.44 (m, 6 H, ArH).
^{13}C–NMR (75 MHz, CDCl$_3$): δ = 55.5 (OMe), 55.5 (CHBr), 114.0 (Ar), 118.4 (Ar), 119.4 (Ar), 123.8 (Ar), 129.8 (Ar), 129.9 (Ar), 130.0 (Ar), 133.5 (Ar), 136.1 (Ar), 156.8 (Ar), 157.3 (Ar), 159.5 (Ar).
MS (+EI): m/z (%) = 289.3 (100) [M$^+$–Br].
m.p.: 61.3 °C - 62.0 °C

15-Br was obtained from 4,4'-dimethoxy benzhydrol (1.31 g, 5.36 mmol) and acetyl bromide (6.61 g, 53.8 mmol). The product was isolated as slightly pink crystals (1.27 g, 77 %). ^{1}H-NMR-spectra are in agreement with literature data.[56]
^{1}H–NMR (400 MHz, CDCl$_3$): δ = 3.80 (s, 6 H, OMe), 6.32 (s, 1 H, CHBr), 6.85-6.87 (m, 4 H, ArH), 7.37-7.39 (m, 4 H, ArH).
^{13}C–NMR (100 MHz, CDCl$_3$): δ = 55.5 (OMe), 56.1 (CHBr), 113.9 (Ar), 129.8 (Ar), 133.7 (1',Ar), 159.3 (Ar).
MS (+EI): m/z (%) = 227.2 (100) [M$^+$–Br].
m.p.: 72.8 °C - 73.9 °C
Elemental Analysis: Calculated for C$_{15}$H$_{15}$BrO$_2$:   C, 58.65; H, 4.92.
                    Found: C, 58.67; H, 4.90.

2. Nucleofugality and Nucleophilicity of Fluoride in Protic Solvents

Preparation of benzhydryl fluorides 11-F and 15-F

As stated above benzhydryl fluorides **11-F** and **15-F** could be isolated. The other benzhydryl fluorides were not isolated and used directly in solution for the solvolytic investigations.

11-F

4,4'-Dimethylbenzhydryl fluoride

In a flame-dried Schlenk flask, 4.85 g (containing 7.28 mmol AgF) of the fluorination agent was suspended in 12 mL of acetonitrile. 4,4'-Dimethylbenzhydryl bromide (5b) (1.00 g, 3.63 mmol) was dissolved in 5 mL of acetonitrile and added dropwise at 0 °C. After stirring for 30 min, the solution was filtered and the solvent was removed under reduced pressure at room temperature. The crude product was distilled in the vacuum (bp. 160 °C/5.0 × 10^{-3} mbar) to give a colorless oil (0.61 g, 78 %).

^{1}H-NMR (400 MHz, CD$_3$CN): δ = 2.33 (s, 6 H, Me), 6.47 (d, 1 H, $^2J_{HF}$ = 44.0 Hz, CHF), 7.19-7.26 (m, 8 H, ArH).

^{13}C-NMR (100 MHz, CD$_3$CN): δ = 21.2 (Me), 95.2 (d, $^1J_{CF}$ = 170.0 Hz, CHF), 127.2 (d, $^3J_{CF}$ = 6.0 Hz), 130.1 (3-Ar), 138.5 (d, $^2J_{CF}$ = 22.1 Hz), 139.3 (d, $^5J_{CF}$ = 2.0 Hz).

^{19}F NMR (282 MHz, CD$_3$CN): δ = -166.2 (d, $^2J_{FH}$ = 45.2 Hz, CHF).

MS (+ EI): m/z (%) = 214.3 (42) [M$^+$], 199.2 (100) [C$_{14}$H$_{12}$F$^+$].

Elemental Analysis: Calculated for C$_{15}$H$_{15}$F: C, 84.08; H, 7.06.
                        Found: C, 84.27; H, 6.78.

15-F

4,4'-Dimethoxybenzhydryl fluoride

In a flame-dried Schlenk flask, 3.91 g (containing 5.86 mmol AgF) of the fluorination agent was suspended in 5 mL of acetonitrile. 4,4'-Dimethoxybenzhydryl bromide (S5) (1.5 g, 4.88 mmol) was dissolved in 10 mL of acetonitrile and added dropwise at 0 °C. After stirring for 30 min, the solution was filtered and the solvent was removed under reduced pressure at room temperature. The crude product was recrystallized from dichloromethane/n-pentane to give slightly pink crystals (0.60 g, 50 %). The product was stored under Argon at −35 °C.

2. Nucleofugality and Nucleophilicity of Fluoride in Protic Solvents

^{1}H-NMR (400 MHz, CD$_3$CN): δ = 3.78 (s, 6 H, OMe), 6.45 (d, 1 H, $^2J_{HF}$ = 48.0 Hz, CHF), 6.92-6.95 (m, 4 H, ArH), 7.26-7.30 (m, 4 H, ArH).
^{13}C-NMR (100 MHz, CD$_3$CN): δ = 56.0 (OMe), 94.9 (d, $^1J_{CF}$ = 169.0 Hz, CHF), 114.8 (Ar), 128.9 (d, $^3J_{CF}$ = 6.0 Hz), 133.4 (d, $^2J_{CF}$ = 23.0 Hz), 160.7 (d, $^5J_{CF}$ = 2.0 Hz).
MS (+ EI): m/z (%) = 246.1 (0.1) [M$^+$], 228.1 (38) [C$_{15}$H$_{16}$O$_2^+$], 227.1 (100) [M$^+$−F].
HRMS (+EI) Calcd. for C$_{15}$H$_{15}$FO$_2$: 246.1056; Found: 246.1045.
T_m = 65.8 °C - 66.5 °C

Procedure for the synthesis of stock solutions of the benzhydryl fluorides (12-14)-F for kinetic measurements.

In a flame-dried Schlenk flask, 1.00 g (containing 1.5 mmol AgF) of the fluorinating agent was suspended in 3 mL of acetonitrile. The corresponding benzhydryl bromide (1.0 mmol) was dissolved in 2 mL of acetonitrile and added to the stirred suspension at 0 °C. After 30 min the solution was filtered. ^{1}H-NMR spectra that were recorded from the resulting solution after evaporation of acetonitrile in deuterated acetonitrile showed the characteristic doublet for CHF. The clear solution is used directly for solvolytic measurements and can be used for approximately 2 days.

Kinetics of Solvolysis Reactions

Solvolysis rate constants of benzhydryl derivatives (Tables 2.2 and 2.9) were monitored by conventional conductometry. Freshly prepared solvents (30 mL) were thermostated (±0.1 °C) at the given temperature for 5 min prior to adding the substrate. Typically 0.25 mL of a 0.2 M stock solution of the substrate in acetonitrile was injected into the solvent. After injection of the benzhydrylium derivative into the solvolyzing medium, an increase of conductance was observed, which was recorded at certain time intervals resulting in about 3000 data points for each measurement. The first-order rate constants k_1 (s^{-1}) were obtained by least squares fitting of the conductance data to a single-exponential equation $G = G_\infty(1-e^{-k_1 t}) + C$. Each rate constant was typically averaged from at least three kinetic runs. Only in two cases (**15-F** in 90A10W and **11-F** in 80E20W) measurements were only performed once as these reactions were very slow. All solvolyses were performed at 25 °C. The following solvents were commercially available and were used as received: Acetone

2. Nucleofugality and Nucleophilicity of Fluoride in Protic Solvents

(99.8 %), acetonitrile (extra dry, water content < 50 ppm), methanol (99.8 %). Dry ethanol was obtained by distillation of commercially available absolute ethanol from sodium/diethyl phthalate. Doubly distilled water [Impendance 18.2 Ω] was prepared with a water purification system.

Calibration experiments were performed by stepwise addition of 50 µL portions of a 0.2 M solution of **14-F** in acetonitrile to 30 mL of 80E20W containing 0.08 M piperidine. After the addition of a portion of **14-F** the conductance after at least 200 s (half-life of **14-F** in 80E20W 49 s) was recorded before the next portion of **14-F** was added. Plots of the conductance against the initial concentration of added benzhydryl fluoride **14-F** were linear (Figure 2.11). Therefore, the solvolytic rate constants can be determined reliably by time-dependent conductance measurements.

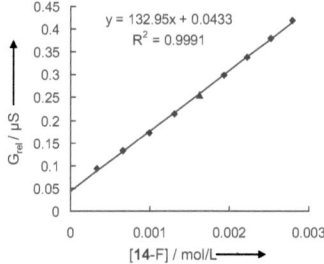

Figure 2.11. Conductance at t_∞ vs. concentration benzhydryl fluoride **14-F** in 80E20W. After the addition of a portion of **14-F**, the next conductance value was taken when the conductance remained constant.

2. Nucleofugality and Nucleophilicity of Fluoride in Protic Solvents

Table 2.9. Individual rate constants for the solvolysis reactions of benzhydryl fluorides **11-15-F**.

solvent [a]	benzhydryl fluoride	[amine]/M [b]	k_1 (individual)/[s^{-1}]	k_1 (average)/[s^{-1}]
90A10W	**15-F**	[pip] = 0.08	9.26×10^{-5}	9.26×10^{-5}
80A20W	**15-F**	[pip] = 0.08	1.24×10^{-3}	1.20×10^{-3}
		[pip] = 0.08	1.18×10^{-3}	
		[pip] = 0.16	1.17×10^{-3}	
	14-F	[pip] = 0.08	1.47×10^{-4}	1.43×10^{-4}
		[pip] = 0.08	1.40×10^{-4}	
		[pip] = 0.16	1.41×10^{-4}	
80AN20W	**15-F**	[pip] = 0.08	7.51×10^{-3}	7.90×10^{-3}
		[pip] = 0.08	8.09×10^{-3}	
		[pip] = 0.16	8.10×10^{-3}	
	14-F	[pip] = 0.08	1.17×10^{-3}	1.11×10^{-3}
		[pip] = 0.08	1.11×10^{-3}	
		[pip] = 0.16	1.06×10^{-3}	
	13-F	[pip] = 0.08	5.05×10^{-4}	4.85×10^{-4}
		[pip] = 0.16	4.66×10^{-4}	
		[pip] = 0.16	4.83×10^{-4}	
100E	**15-F**	[pip] = 0.08	4.37×10^{-3}	4.26×10^{-3}
		[pip] = 0.12	4.27×10^{-3}	
		[pip] = 0.16	4.13×10^{-3}	
	14-F	[pip] = 0.08	5.89×10^{-4}	5.63×10^{-4}
		[pip] = 0.08	5.39×10^{-4}	
		[pip] = 0.16	5.89×10^{-4}	
		[pip] = 0.16	5.34×10^{-4}	

2. Nucleofugality and Nucleophilicity of Fluoride in Protic Solvents

Table 2.9. (continued)

solvent [a]	benzhydryl fluoride	[amine]/M [b]	k_1 (individual)/[s^{-1}]	k_1 (average)/[s^{-1}]
100E	13-F	[pip] = 0.08	1.66×10^{-4}	1.63×10^{-4}
		[pip] = 0.08	1.63×10^{-4}	
		[lut] = 0.09	1.64×10^{-4}	
		[pip] = 0.16	1.57×10^{-4}	
60A40W	14-F	[pip] = 0.08	3.55×10^{-3}	3.55×10^{-3}
		[pip] = 0.08	3.65×10^{-3}	
		[pip] = 0.08	3.66×10^{-3}	
		[pip] = 0.12	3.55×10^{-3}	
		[pip] = 0.16	3.35×10^{-3}	
	13-F	[pip] = 0.08	1.89×10^{-3}	1.87×10^{-3}
		[pip] = 0.08	1.81×10^{-3}	
		[pip] = 0.08	1.93×10^{-3}	
		[pip] = 0.16	1.85×10^{-3}	
	12-F	[pip] = 0.08	3.84×10^{-4}	3.80×10^{-4}
		[pip] = 0.08	3.84×10^{-4}	
		[pip] = 0.12	3.71×10^{-4}	
	11-F	[pip] = 0.08	3.39×10^{-5}	3.28×10^{-5}
		[pip] = 0.08	3.12×10^{-5}	
		[pip] = 0.08	3.32×10^{-5}	
100M	15-F	[pip] = 0.08	3.95×10^{-2}	3.92×10^{-2}
		[pip] = 0.08	3.91×10^{-2}	
		[pip] = 0.08	3.87×10^{-2}	
		[pip] = 0.08	3.91×10^{-2}	
		[pip] = 0.08	3.95×10^{-2}	
	14-F	[pip] = 0.08	5.71×10^{-3}	5.49×10^{-3}
		[pip] = 0.12	5.47×10^{-3}	
		[pip] = 0.12	5.30×10^{-3}	

2. Nucleofugality and Nucleophilicity of Fluoride in Protic Solvents

Table 2.9. (continued)

solvent [a]	benzhydryl fluoride	[amine]/M [b]	k_1 (individual)/[s^{-1}]	k_1 (average)/[s^{-1}]
100M	13-F	[pip] = 0.08	2.15×10^{-3}	1.99×10^{-3}
		[pip] = 0.08	2.07×10^{-3}	
		[pip] = 0.16	1.88×10^{-3}	
		[pip] = 0.20	1.88×10^{-3}	
	12-F	[pip] = 0.08	3.52×10^{-4}	3.43×10^{-4}
		[pip] = 0.08	3.45×10^{-4}	
		[pip] = 0.16	3.30×10^{-4}	
80E20W	15-F	[pip] = 0.08	7.81×10^{-2}	7.77×10^{-2}
		[pip] = 0.08	7.71×10^{-2}	
		[pip] = 0.08	7.60×10^{-2}	
		[pip] = 0.08	7.98×10^{-2}	
	14-F	[pip] = 0.08	1.40×10^{-2}	1.42×10^{-2}
		[pip] = 0.16	1.40×10^{-2}	
		[pip] = 0.16	1.46×10^{-2}	
	13-F	[pip] = 0.08	4.37×10^{-3}	4.34×10^{-3}
		[pip] = 0.16	4.33×10^{-3}	
		[pip] = 0.16	4.31×10^{-3}	
	12-F	[pip] = 0.08	8.04×10^{-4}	7.82×10^{-4}
		[pip] = 0.08	7.91×10^{-4}	
		[pip] = 0.16	7.50×10^{-4}	
	11-F	[pip] = 0.08	5.75×10^{-5}	5.75×10^{-5}
60AN40W	15-F	[pip] = 0.08	8.28×10^{-2}	8.35×10^{-2}
		[pip] = 0.08	8.50×10^{-2}	
		[pip] = 0.08	8.58×10^{-2}	
		[pip] = 0.08	8.02×10^{-2}	

2. Nucleofugality and Nucleophilicity of Fluoride in Protic Solvents

Table 2.9. (continued)

solvent [a]	benzhydryl fluoride	[amine]/M [b]	k_1 (individual)/[s^{-1}]	k_1 (average)/[s^{-1}]
60AN40W	14-F	[pip] = 0.08	9.49×10^{-3}	9.28×10^{-3}
		[pip] = 0.11	9.00×10^{-3}	
		[lut] = 0.12	9.62×10^{-3}	
		[pip] = 0.16	9.01×10^{-3}	
	13-F	[pip] = 0.08 [c]	4.81×10^{-3}	4.66×10^{-3}
		[pip] = 0.08 [c]	4.70×10^{-3}	
		[pip] = 0.16 [c]	4.47×10^{-3}	
		[pip] = 0.11	4.66×10^{-3}	
		[pip] = 0.19	4.41×10^{-3}	
		[pip] = 0.21	4.54×10^{-3}	
		[pip] = 0.33	4.01×10^{-3}	
		[pip] = 0.41	4.12×10^{-3}	
		[lut] = 0.09	4.54×10^{-3}	
		[lut] = 0.15	4.71×10^{-3}	
		[lut] = 0.23	4.41×10^{-3}	
		[NEt$_3$] = 0.08	4.41×10^{-3}	
		[NEt$_3$] = 0.11	4.66×10^{-3}	
		[NEt$_3$] = 0.15	4.71×10^{-3}	
	12-F	[pip] = 0.08	9.40×10^{-4}	1.02×10^{-3}
		[pip] = 0.08	9.37×10^{-4}	
		[pip] = 0.08	1.14×10^{-3}	
		[pip] = 0.08	1.04×10^{-3}	

2. Nucleofugality and Nucleophilicity of Fluoride in Protic Solvents

Table 2.9. (continued)

solvent [a]	benzhydryl fluoride	[amine]/M [b]	k_1 (individual)/[s^{-1}]	k_1 (average)/[s^{-1}]
	11-F	[pip] = 0.08	9.48 × 10^{-5}	9.99 × 10^{-5}
		[pip] = 0.08	1.03 × 10^{-4}	
		[pip] = 0.08	1.02 × 10^{-4}	
		[pip] = 0.08	9.79 × 10^{-5}	
		[pip] = 0.08	9.79 × 10^{-5}	
		[pip] = 0.08	9.59 × 10^{-5}	
		[pip] = 0.16	1.05 × 10^{-4}	

[a] Mixtures of solvents are given as (v/v); solvents: A = acetone, AN = acetonitrile, E = ethanol, M = methanol, W = water. [b] pip = piperidine, lut = 2,6-lutidine [c] only entries 1-3 were used for the calculation of the average k_1 for **13-F** in 60AN40W the other rate constants were used in Figure 2.1.

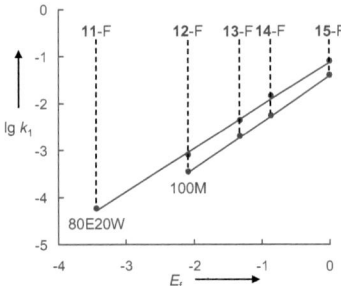

Figure 2.12. Plot of lg k_1 for the solvolysis reactions of various benzhydryl fluorides vs. electrofugalities E_f (systems not depicted in Figure 2.2): in 80E20W (v/v) and 100M. E= ethanol, M= methanol, W= water.

2. Nucleofugality and Nucleophilicity of Fluoride in Protic Solvents

Determination of the Eyring and Arrhenius activation parameters

Table 2.10. Rate constants k_1 of the solvolysis reaction of various benzhydryl halides in 100E, and 60AN40W at different temperatures.

T/°C	k_1/s^{-1} 15-F in 100E	T/°C	k_1/s^{-1} 13-F in 60AN40W
40.6	1.45×10^{-2}	40.0	1.64×10^{-2}
31.6	7.14×10^{-3}	25.0	4.66×10^{-3}
25.0	4.26×10^{-3}	16.1	2.04×10^{-3}
13.8	1.47×10^{-3}	8.1	8.97×10^{-4}
5.2	6.12×10^{-4}		

	7-Cl in 60AN40W		7-Br in 60AN40W
66.0	1.78×10^{-2}	35.0	1.50×10^{-2}
56.0	7.31×10^{-3}	25.0	4.59×10^{-3}
45.0	2.46×10^{-3}	16.1	1.60×10^{-3}
35.0	9.14×10^{-4}	6.4	4.24×10^{-4}
25.0	2.67×10^{-4}		

Table 2.11. Activation parameters for the solvolyses of various benzhydryl halides in 100E and 60AN40W.

	15-F in 100E	13-F in 60AN40W	7-Cl in 60AN40W	7-Br in 60AN40W
$\Delta H^{\ddagger}$/kJ mol^{-1}	62.5 ± 1.0	64.2 ± 0.7	83.0 ± 1.2	86.5 ± 1.0
$\Delta S^{\ddagger}$/J mol^{-1} K^{-1}	−81.0 ± 3.3	−74.5 ± 2.4	−34.6 ± 3.9	0.68 ± 3.3
E_a/kJ mol^{-1}	65.0 ± 1.0	66.6 ± 0.7	85.6 ± 1.2	89.0 ± 1.0
lg A	8.99 ± 0.17	9.33 ± 0.12	11.4 ± 0.20	13.3 ± 0.17

2. Nucleofugality and Nucleophilicity of Fluoride in Protic Solvents

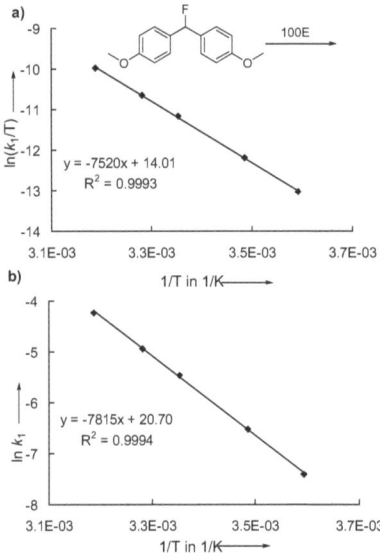

Figure 2.13. Eyring (a) and Arrhenius (b) plot for **15**-F in 100E.

2. Nucleofugality and Nucleophilicity of Fluoride in Protic Solvents

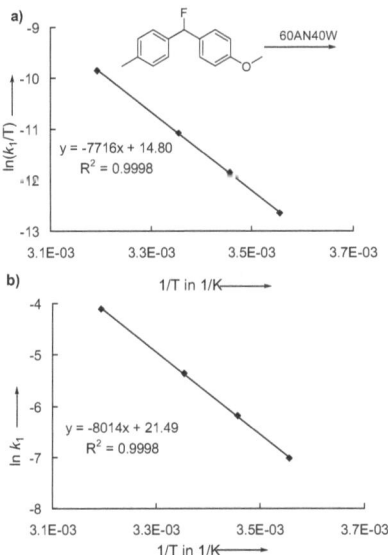

Figure 2.14. Eyring (a) and Arrhenius (b) plot for **13-F** in 60AN40W.

2. Nucleofugality and Nucleophilicity of Fluoride in Protic Solvents

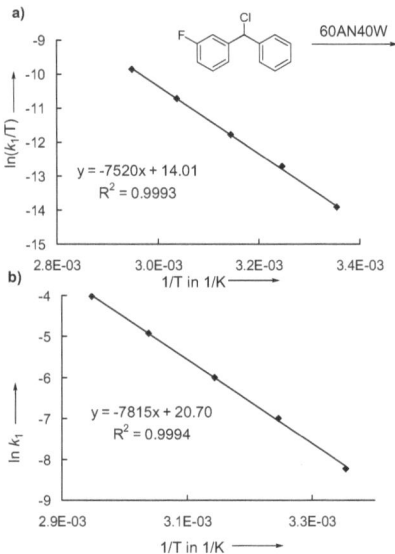

Figure 2.15. Eyring (a) and Arrhenius (b) plot for **7-Cl** in 60AN40W.

2. Nucleofugality and Nucleophilicity of Fluoride in Protic Solvents

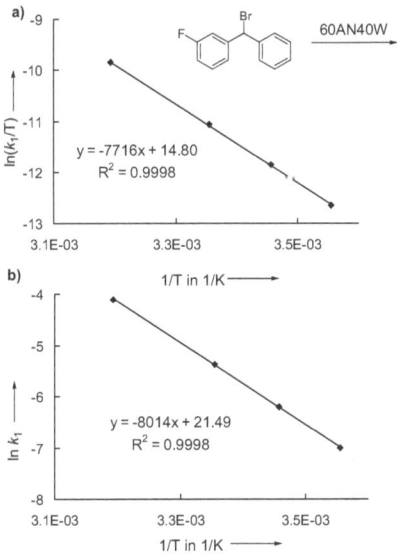

Figure 2.16. Eyring (a) and Arrhenius (b) plot for **7-Br** in 60AN40W.

2. Nucleofugality and Nucleophilicity of Fluoride in Protic Solvents

Determination of the Nucleophilicity of Fluoride and k_{solv}

Kinetic method

Solvents and fluoride sources:
Potassium fluoride of p.a. grade, cesium fluoride (99.9 %) and tetrabutylammonium fluoride trihydrate (98 %) were used as fluoride sources for the kinetic experiments. Acetonitrile (extra dry, water content < 50 ppm), methanol (99.8 %) were used without further purification for laser-flash experiments. Dry ethanol was obtained by distillation of commercially available absolute ethanol from sodium/diethyl phthalate. Doubly distilled water [Impedance 18.2 Ω] was prepared with a water purification system.

Laser-flash experiments:
The measurements were performed in a room with 20 ± 1 °C, regulated by a clima control unit. Samples were stored sufficiently long in this room to exhibit a temperature of 20 ± 1 °C. The benzhydryl cations **11$^+$-19$^+$** were generated by irradiating solutions ($A_{266\,nm} \approx 0.1$-1.0) of the precursor salts **11-17**-PPh$_3$ or **18,19**-PBu$_3$ with a 7-ns laser pulse (7 ns pulse width, 266 nm, 40-60 mJ/pulse). The benzhydryl cations **11$^+$-17$^+$** were generated from the triphenylphosphonium salts **11-17**-PPh$_3$. For the photogeneration of **18$^+$,19$^+$**, stock solutions of the tributylphosphonium salts **18,19**-PBu$_3$ in acetonitrile were prepared by mixing the appropriate amounts of benzhydrylium tetrafluoroborates **11$^+$-17$^+$**-BF$_4^-$ and tributylphosphine. The system was equipped with a fluorescence flow cell which allowed complete replacing of the sample volume between subsequent laser pulses. Kinetics were measured by following the UV-vis absorbance decay of the benzhydrylium cations at their absorbance maxima. Averaged data obtained from ≥ 48 individual runs were used for further evaluations. First-order rate constants (k_{obs}) were calculated by least-squares fitting of the absorbance data to a single exponential function $A_t = A_0\,e^{(-k_{obs}t)} + C$. The second-order rate constants (k_1) were obtained using the slopes of the linear plots of k_{obs} against the nucleophile concentration ([F$^-$]). First-order rate constants (k_{solv}) were obtained from independent measurements with laser-photolytic generated benzhydrylium ions in the absence of fluoride.

2. Nucleofugality and Nucleophilicity of Fluoride in Protic Solvents

Stopped-flow experiments:
The kinetics of the reactions of F^- with 20^+ and 21^+ were determined by the stopped-flow method using the isolated benzhydrylium tetrafluoroborates as described previously.[37] The decay of their absorbance was monitored by UV-vis spectroscopy at their absorption maxima. Pseudo-first-order rate constants k_{obs} were obtained from at least six runs at each fluoride concentration. The absorbance-time curves were fitted to the single exponential function, $A_t = A_0\, e^{(-k_{obs}t)} + C$ to yield the rate constants k_{obs} (s^{-1}). The second-order rate constants k_{-1} ($M^{-1}\,s^{-1}$) for the reactions of 20^+ and 21^+ with fluoride were obtained from the slopes of plots of k_{obs} versus the fluoride concentrations.

Product study:
For product analysis 4,4'-dimethylbenzhydryl chloride (**11-Cl**; 25 mg 0.11 mmol) was dissolved in 0.2 mL of deuterated acetonitrile and transferred to a NMR-tube. Possible reaction pathways are shown in Scheme 2.4 Then tetrabutylammonium fluoride trihydrate (120 mg, 0.39 mmol) in a mixture of 0.22 mL deuterated acetonitrile and 0.28 mL D_2O was added rapidly, and the NMR tube was shaken intensively to ensure fast mixing. Immediately afterwards an NMR-spectrum was recorded.

As stated in chapter 2.2 a more precise mathematical treatment of the kinetic results and the ^{1}H-NMR experiment is provided here. According to the calculated rate constant of $k_1 = 2.42\,s^{-1}$, tol$_2$CHCl will ionize almost instantly in 60AN40W yielding the benzhydrylium ion 11^+. The rate constants for the reaction of 11^+ with 60 % aqueous acetonitrile (k_{solv}) and fluoride in acetonitrile (k_{-1}) have been determined (see below).

Scheme 2.4. Possible reaction pathways for the reaction of (4-MeC$_6$H$_4$)$_2$CHCl (**11-Cl**) in 60 % aqueous acetonitrile containing fluoride.

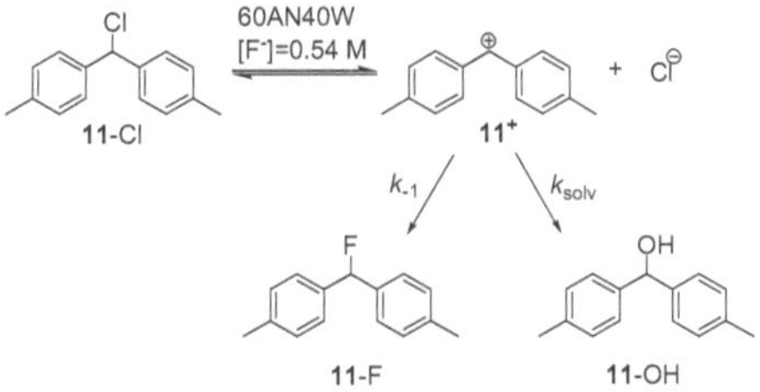

2. Nucleofugality and Nucleophilicity of Fluoride in Protic Solvents

According to Huisgen,[58] the competition constant can be calculated by equation 2.7.

$$\chi = \frac{\log[F^-]_0 - \log([F^-]_0 - [11-F]_e)}{\log[H_2O]_0 - \log([H_2O]_0 - [11-OH]_e)} \quad (2.7)$$

From the initial concentration of **11-Cl** (tol$_2$CHCl) and the product ratios from the ^{1}H-NMR spectra recorded after mixing a solution of **11-Cl** in acetonitrile with an aqueous solution of tetrabutylammonium fluoride trihydrate the final concentrations ([11-F]$_e$ and [11-OH]$_e$ of **11-F** and **11-OH** were calculated.

$$[11-F]_e = [11-Cl]_i \times \frac{2.36}{3.36} = 0.11 \text{ M}$$

$$[11-OH]_e = [11-Cl]_i \times \frac{1.00}{3.36} = 4.6 \times 10^{-2} \text{ M}$$

Thus, a competition constant of $\chi = 108$ is calculated from the product analysis by NMR experiment. This value is similar to that calculated from the rate constants given in Table 2.4 (eq. 2.8).

$$\chi = \frac{k_{-1}}{k_{solv}/[H_2O]} = \frac{2.0 \times 10^8}{3.35 \times 10^7 / 22.2} = 133 \quad (2.8)$$

Kinetics of the reactions of benzhydrylium ions with solvents:

Table 2.10. Rate constants k_{solv} for the reaction of various benzhydrylium ions with pure solvents.

	90AN10W		
	[N°-PPh$_3$]/M	λ/nm	k_{solv}/s^{-1}
17$^+$	4.15 × 10^{-5}	523	7.05 × 10$^{3\,a}$
16$^+$	4.15 × 10^{-5}	513	2.12 × 10^4
15$^+$	4.67 × 10^{-5}	500	6.78 × 10$^{4\,b}$

^{a} A value of 7.11 × 10^2 was reported in Ref. [59] ^{b} A value of 9.87 × 10^4 was reported in Ref. [59]

2. Nucleofugality and Nucleophilicity of Fluoride in Protic Solvents

Table 2.10. Rate constants k_{solv} for the reaction of various benzhydrylium ions with pure solvents.

	[N°-PPh₃]/M	λ/nm	k_{solv}/s^{-1}
	100M		
17⁺	9.39×10^{-6}	528	7.39×10^{5}
16⁺	2.32×10^{-5}	513	2.44×10^{6}
14⁺	9.72×10^{-6}	509	1.97×10^{7}
	98AN2W		
12⁺	4.52×10^{-5}	455	3.85×10^{5}
11⁺	2.38×10^{-5}	464	6.18×10^{6}
	80AN20W		
17⁺	2.03×10^{-5}	523	9.38×10^{3}
16⁺	2.20×10^{-5}	513	3.08×10^{4}
15⁺	2.24×10^{-5}	500	$9.49 \times 10^{4\,b}$
14⁺	2.19×10^{-5}	500	2.52×10^{5}
	60AN40W		
17⁺	4.60×10^{-5}	523	1.14×10^{4}
16⁺	2.26×10^{-5}	513	3.16×10^{4}
15⁺	6.02×10^{-5}	500	9.61×10^{4}
14⁺	2.51×10^{-5}	500	2.81×10^{5}
13⁺	5.03×10^{-5}	478	1.03×10^{6}
11⁺	1.46×10^{-4}	464	3.35×10^{7}
	80E20W		
16⁺	2.41×10^{-5}	513	4.87×10^{5}
15⁺	4.78×10^{-5}	500	$1.66 \times 10^{6\,c}$

[b] A value of 9.82×10^{4} was reported in Ref. [59] [c] A value of 1.51×10^{6} was reported in Ref. [59]

2. Nucleofugality and Nucleophilicity of Fluoride in Protic Solvents

Kinetics of the reactions of benzhydrylium ions with fluoride:

Table 2.11. Rate constants k_{obs} for the reaction of various benzhydrylium ions with fluoride in 90AN10W.

[17-PPh$_3$] / M	[Bu$_4$N$^+$F$^-$ × 3 H$_2$O]/M	k_{obs}/s^{-1}	λ = 523 nm	k_{-1}/M^{-1} s^{-1}
4.15 × 10^{-5}	3.10 × 10^{-2}	7.54 × 10^4		2.89 × 10^6
	6.66 × 10^{-2}	1.44 × 10^5		
	8.95 × 10^{-2}	2.46 × 10^5		
	1.03 × 10^{-1}	2.71 × 10^5		
	1.33 × 10^{-1}	3.63 × 10^5		

[16-PPh$_3$] / M	[Bu$_4$N$^+$F$^-$ × 3 H$_2$O]/M	k_{obs}/s^{-1}	λ = 513 nm	k_{-1}/M^{-1} s^{-1}
4.15 × 10^{-5}	4.21 × 10^{-2}	2.64 × 10^5		6.03 × 10^6
	5.76 × 10^{-2}	3.32 × 10^5		
	8.53 × 10^{-2}	5.03 × 10^5		
	1.13 × 10^{-1}	6.86 × 10^5		

[15-PPh$_3$] / M	[Bu$_4$N$^+$F$^-$ × 3 H$_2$O]/M	k_{obs}/s^{-1}	λ = 500 nm	k_{-1}/M^{-1} s^{-1}
4.67 × 10^{-5}	2.46 × 10^{-2}	4.65 × 10^5		1.85 × 10^7
	5.52 × 10^{-2}	9.51 × 10^5		
	7.29 × 10^{-2}	1.32 × 10^6		
	1.07 × 10^{-1}	1.87 × 10^6		
	1.26 × 10^{-1}	2.38 × 10^6		

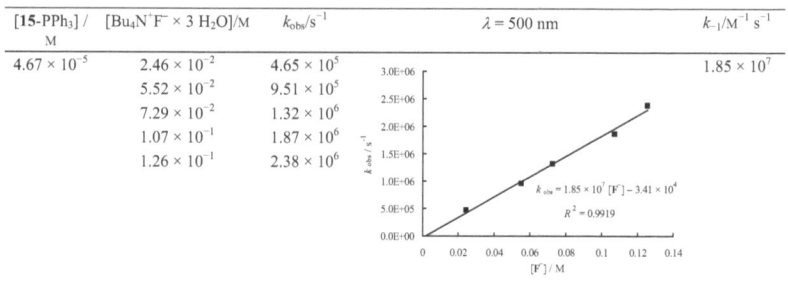

2. Nucleofugality and Nucleophilicity of Fluoride in Protic Solvents

Table 2.12. Rate constants k_{obs} for the reaction of various benzhydrylium ions with fluoride in 100M.

[**17**-PPh$_3$] / M	[K$^+$F$^-$]/M	k_{obs}/s^{-1}	λ = 523 nm	k_{-1}/M^{-1} s^{-1}
6.04 × 10^{-5}	7.63 × 10^{-2}	8.06 × 10^5		1.56 × 10^6
	1.47 × 10^{-1}	9.81 × 10^5		
	2.59 × 10^{-1}	1.15 × 10^6	$k_{obs} = 1.56 \times 10^6$ [F$^-$] + 7.16 × 10^5	
	3.37 × 10^{-1}	1.21 × 10^6	$R^2 = 0.9965$	
	5.79 × 10^{-1}	1.61 × 10^6		
	8.93 × 10^{-1}	2.12 × 10^6		

[**16**-PPh$_3$] / M	[K$^+$F$^-$]/M	k_{obs}/s^{-1}	λ = 513 nm	k_{-1}/M^{-1} s^{-1}
6.56 × 10^{-5}	6.55 × 10^{-2}	2.48 × 10^6		4.12 × 10^6
	1.28 × 10^{-1}	2.75 × 10^6		
	2.26 × 10^{-1}	3.24 × 10^6	$k_{obs} = 4.12 \times 10^6$ [F$^-$] + 2.24 × 10^6	
	3.76 × 10^{-1}	3.79 × 10^6	$R^2 = 0.9977$	
	6.15 × 10^{-1}	4.70 × 10^6		
	7.19 × 10^{-1}	5.25 × 10^6		

[**15**-PPh$_3$] / M	[K$^+$F$^-$]/M (■)	k_{obs}/s^{-1}	λ = 500 nm	k_{-1}/M^{-1} s^{-1}
3.87 × 10^{-5}	1.12 × 10^{-1}	1.11 × 10^7		1.14 × 10^7
	1.48 × 10^{-1}	1.17 × 10^7		
	2.52 × 10^{-1}	1.31 × 10^7		
	3.17 × 10^{-1}	1.36 × 10^7		
	4.74 × 10^{-1}	1.58 × 10^7		
	7.09 × 10^{-1}	1.77 × 10^7		

[**15**-PPh$_3$] / M	[Cs$^+$F$^-$]/M (▲)	k_{obs}/s^{-1}		
1.93 × 10^{-5}	3.09 × 10^{-2}	9.61 × 10^6	$k_{obs} = 1.14 \times 10^7$ [F$^-$] + 9.91 × 10^6	
	7.63 × 10^{-2}	1.05 × 10^7	$R^2 = 0.966$	
	1.15 × 10^{-1}	1.14 × 10^7		
	2.84 × 10^{-1}	1.38 × 10^7		
	2.95 × 10^{-1}	1.35 × 10^7		
	5.02 × 10^{-1}	1.49 × 10^7		

2. Nucleofugality and Nucleophilicity of Fluoride in Protic Solvents

Table 2.13. Rate constants k_{obs} for the reaction of various benzhydrylium ions with fluoride in 98AN2W

[21$^+$]/M	[Bu$_4$N$^+$F$^-$ × 3 H$_2$O] M	k_{obs}/s^{-1}	λ = 613 nm	k_{-1}/M^{-1} s^{-1}
2.23 × 10^{-5}	3.09 × 10^{-4}	2.67		1.97 × 10$^{3\,a}$
	9.28 × 10^{-4}	4.01	k_{obs} = 1.97 × 10^3 [F$^-$] + 2.11 × 10^0	
	1.24 × 10^{-3}	4.56	R^2 = 0.9966	
	1.55 × 10^{-3}	5.10		

[20$^+$]/M	[Bu$_4$N$^+$F$^-$ × 3 H$_2$O]/M	k_{obs}/s^{-1}	λ = 620 nm	k_{-1}/M^{-1} s^{-1}
3.58 × 10^{-5}	1.10 × 10^{-3}	4.76 × 10^1		2.95 × 10$^{4\,a}$
	6.28 × 10^{-4}	4.07 × 10^1		
	1.57 × 10^{-3}	6.96 × 10^1	k_{obs} = 2.95 × 10^4 [F$^-$] + 1.96 × 10^1	
	2.04 × 10^{-3}	7.59 × 10^1	R^2 = 0.9696	
	2.51 × 10^{-3}	9.60 × 10^1		

[19-PBu$_3$] / M	[Bu$_4$N$^+$F$^-$ × 3 H$_2$O]/M	k_{obs}/s^{-1}	λ = 586 nm	k_{-1}/M^{-1} s^{-1}
2.73 × 10^{-5}	3.16 × 10^{-3}	2.27 × 10^3		4.17 × 10^5
	5.21 × 10^{-3}	2.47 × 10^3		
	9.02 × 10^{-3}	4.84 × 10^3		
	1.45 × 10^{-2}	6.40 × 10^3	k_{obs} = 4.17 × 10^5 [F$^-$] + 7.24 × 10^2	
	1.72 × 10^{-2}	8.12 × 10^3	R^2 = 0.9782	

^{a} Kinetics of the reactions of **21$^+$** and **20$^+$** with F$^-$ in 98AN2W were determined UV/vis-spectroscopically by rapid mixing of equal amounts of a solution of Bu$_4$N$^+$F$^-$ × 3 H$_2$O in aqueous AN (water content adjusted to obtain 96AN4W) and a solution of **1** in 100AN using the stopped-flow method.

2. Nucleofugality and Nucleophilicity of Fluoride in Protic Solvents

Table 2.13. (continued)

[**18**-PBu$_3$] / M	[Bu$_4$N$^+$F$^-$ × 3 H$_2$O]/M	k_{obs}/s^{-1}	λ = 592 nm	k_{-1}/M^{-1} s^{-1}
2.06 × 10^{-5}	2.36 × 10^{-3}	1.54 × 10^4		2.31 × 10^6
	6.09 × 10^{-3}	2.32 × 10^4		
	8.80 × 10^{-3}	2.84 × 10^4		
	1.32 × 10^{-2} [a]	3.52 × 10^4		
	1.66 × 10^{-2}	5.06 × 10^4		

[**17**-PPh$_3$] / M	[Bu$_4$N$^+$F$^-$ × 3 H$_2$O]/M	k_{obs}/s^{-1}	λ = 523 nm	k_{-1}/M^{-1} s^{-1}
2.18 × 10^{-5}	4.40 × 10^{-3}	7.63 × 10^5		1.12 × 10^8
	1.15 × 10^{-2}	1.65 × 10^6		
	1.71 × 10^{-2}	2.44 × 10^6		
	2.10 × 10^{-2}	2.64 × 10^6		
	2.71 × 10^{-2}	3.53 × 10^6		
	3.73 × 10^{-2}	4.46 × 10^6		

[**16**-PPh$_3$] / M	[Bu$_4$N$^+$F$^-$ × 3 H$_2$O]/M	k_{obs}/s^{-1}	λ = 513 nm	k_{-1}/M^{-1} s^{-1}
2.25 × 10^{-5}	1.22 × 10^{-2}	2.60 × 10^6		1.40 × 10^8
	2.22 × 10^{-2}	4.03 × 10^6		
	3.05 × 10^{-2}	5.27 × 10^6		
	6.35 × 10^{-2}	9.91 × 10^6		
	9.08 × 10^{-2}	1.36 × 10^7		

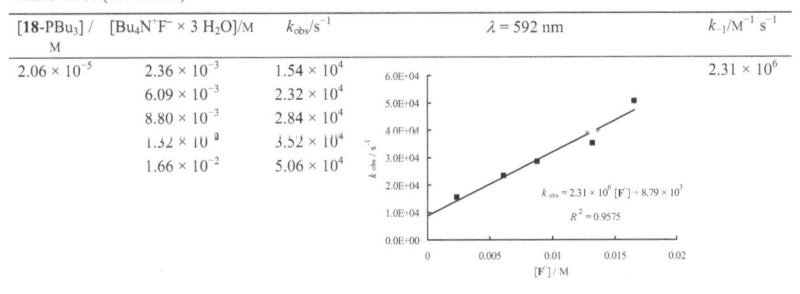

Table 2.13. (continued)

[15-PPh$_3$] / M	[Bu$_4$N$^+$F$^-$ × 3 H$_2$O]/M	k_{obs}/s^{-1}	λ = 500 nm	k_{-1}/M^{-1} s^{-1}
2.78 × 10^{-5}	8.04 × 10^{-3}	4.50 × 10^6		2.69 × 10^8
	1.92 × 10^{-2}	7.25 × 10^6		
	4.29 × 10^{-2}	1.43 × 10^7		
	8.51 × 10^{-2}	2.51 × 10^7		

$k_{obs} = 2.69 \times 10^8$ [F$^-$] + 2.34 × 10^6
$R^2 = 0.9991$

[12-PPh$_3$] / M	[Bu$_4$N$^+$F$^-$ × 3 H$_2$O]/M	k_{obs}/s^{-1}	λ = 455 nm	k_{-1}/M^{-1} s^{-1}
4.52 × 10^{-5}	2.76 × 10^{-3}	9.03 × 10^6		2.40 × 10^9
	5.16 × 10^{-3}	1.47 × 10^7		
	6.85 × 10^{-3}	1.83 × 10^7		
	9.19 × 10^{-3}	2.40 × 10^7		
	2.12 × 10^{-2}	5.30 × 10^7		

$k_{obs} = 2.40 \times 10^9$ [F$^-$] + 2.19 × 10^6
$R^2 = 0.9999$

[11-PPh$_3$] / M	[Bu$_4$N$^+$F$^-$ × 3 H$_2$O]/M	k_{obs}/s^{-1}	λ = 464 nm	k_{-1}/M^{-1} s^{-1}
2.38 × 10^{-5}	9.50 × 10^{-4}	1.89 × 10^7		8.63 × 10^9
	1.90 × 10^{-3}	2.97 × 10^7		
	2.79 × 10^{-3}	3.01 × 10^7		
	3.80 × 10^{-3}	4.56 × 10^7		
	5.24 × 10^{-3}	5.60 × 10^7		

$k_{obs} = 8.63 \times 10^9$ [F$^-$] + 1.07 × 10^7
$R^2 = 0.9619$

2. Nucleofugality and Nucleophilicity of Fluoride in Protic Solvents

Table 2.14. Rate constants k_{obs} for the reaction of various benzhydrylium ions with fluoride in 80AN20W

[17-PPh$_3$] / M	[Bu$_4$N$^+$F$^-$ × 3 H$_2$O]/M	k_{obs}/s^{-1}	λ = 523 nm	k_{-1}/M^{-1} s^{-1}
2.03 × 10^{-5}	3.55 × 10^{-2}	3.06 × 10^4		6.72× 10^5
	7.28 × 10^{-2}	5.01 × 10^4		
	1.12 × 10^{-1}	7.31 × 10^4		
	1.62 × 10^{-1}	1.09 × 10^5		
	1.92 × 10^{-1}	1.36 × 10^5		

[16-PPh$_3$] / M	[Bu$_4$N$^+$F$^-$ × 3 H$_2$O]/M	k_{obs}/s^{-1}	λ = 513 nm	k_{-1}/M^{-1} s^{-1}
2.20 × 10^{-5}	1.81 × 10^{-2}	6.24 × 10^4		1.72 × 10^6
	4.62 × 10^{-2}	1.01 × 10^5		
	6.30 × 10^{-2}	1.26 × 10^5		
	9.89 × 10^{-2}	2.01 × 10^5		

[15-PPh$_3$] / M	[Bu$_4$N$^+$F$^-$ × 3 H$_2$O]/M	k_{obs}/s^{-1}	λ = 500 nm	k_{-1}/ M^{-1} s^{-1}
2.24 × 10^{-5}	4.45 × 10^{-2}	3.03 × 10^5		5.10 × 10^6
	7.39 × 10^{-2}	4.76 × 10^5		
	1.05 × 10^{-1}	5.82 × 10^5		
	1.52 × 10^{-1}	8.63 × 10^5		

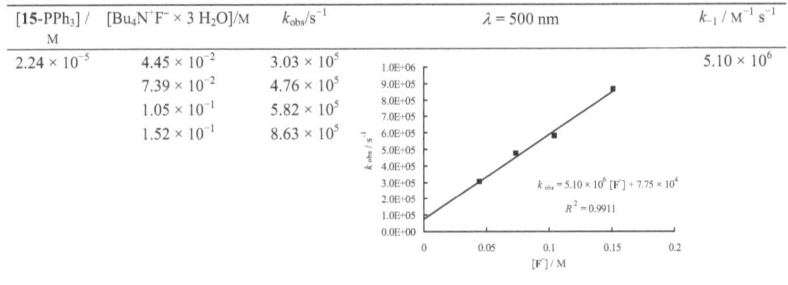

2. Nucleofugality and Nucleophilicity of Fluoride in Protic Solvents

Table 2.14. (continued)

[14-PPh$_3$] / M	[Bu$_4$N$^+$F$^-$ × 3 H$_2$O]/M	k_{obs}/s^{-1}	λ = 500 nm	k_{-1} / M^{-1} s^{-1}
2.19 × 10^{-5}	1.95 × 10^{-2}	4.54 × 10^5		9.76 × 10^6
	4.22 × 10^{-2}	6.36 × 10^5		
	7.24 × 10^{-2}	9.89 × 10^5		
	1.05 × 10^{-1}	1.26 × 10^6		

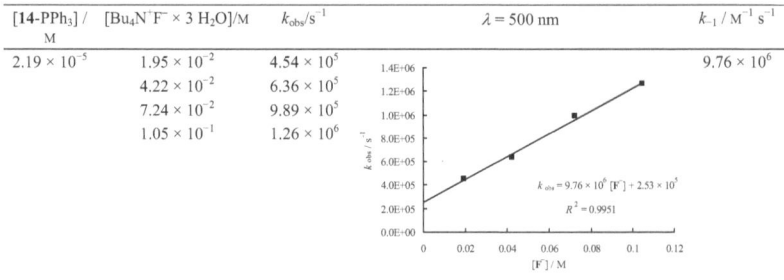

$k_{obs} = 9.76 \times 10^6 \, [\text{F}^-] + 2.53 \times 10^5$
$R^2 = 0.9951$

Table 2.15. Rate constants k_{obs} for the reaction of various benzhydrylium ions with fluoride in 60AN40W

[17-PPh$_3$] / M	[Bu$_4$N$^+$F$^-$ × 3 H$_2$O]/M	k_{obs}/s^{-1}	λ = 523 nm	k_{-1}/M^{-1} s^{-1}
4.60 × 10^{-5}	4.32 × 10^{-2}	1.87 × 10^4		1.66 × 10^5
	8.67 × 10^{-2}	2.47 × 10^4		
	1.64 × 10^{-1}	3.58 × 10^4		
	2.15 × 10^{-1}	4.42 × 10^4		
	2.68 × 10^{-1}	5.72 × 10^4		

$k_{obs} = 1.66 \times 10^5 \, [\text{F}^-] + 1.03 \times 10^4$
$R^2 = 0.9859$

2. Nucleofugality and Nucleophilicity of Fluoride in Protic Solvents

Table 2.15. (continued)

[16-PPh₃] / M	[Bu₄N⁺F⁻ × 3 H₂O]/M	k_{obs}/s⁻¹	$\lambda = 513$ nm	k_{-1}/M⁻¹ s⁻¹
2.26×10^{-5}	3.38×10^{-2}	4.52×10^{4}		3.94×10^{5}
	6.01×10^{-2}	5.44×10^{4}		
	1.17×10^{-1}	7.20×10^{4}		
	1.84×10^{-1}	9.95×10^{4}		
	2.26×10^{-1}	1.23×10^{5}		

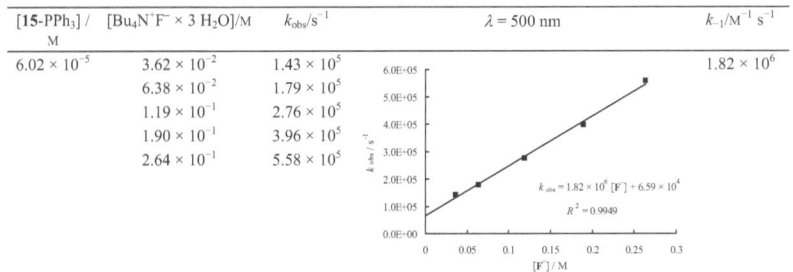

[15-PPh₃] / M	[Bu₄N⁺F⁻ × 3 H₂O]/M	k_{obs}/s⁻¹	$\lambda = 500$ nm	k_{-1}/M⁻¹ s⁻¹
6.02×10^{-5}	3.62×10^{-2}	1.43×10^{5}		1.82×10^{6}
	6.38×10^{-2}	1.79×10^{5}		
	1.19×10^{-1}	2.76×10^{5}		
	1.90×10^{-1}	3.96×10^{5}		
	2.64×10^{-1}	5.58×10^{5}		

[14-PPh₃] / M	[Bu₄N⁺F⁻ × 3 H₂O]/M	k_{obs}/s⁻¹	$\lambda = 500$ nm	k_{-1}/M⁻¹ s⁻¹
2.51×10^{-5}	3.05×10^{-2}	3.80×10^{5}		4.36×10^{6}
	7.23×10^{-2}	5.68×10^{5}		
	1.23×10^{-1}	7.60×10^{5}		
	1.77×10^{-1}	9.78×10^{5}		
	2.07×10^{-1}	1.18×10^{6}		

2. Nucleofugality and Nucleophilicity of Fluoride in Protic Solvents

Table 2.15. (continued)

[13-PPh$_3$] / M	[Bu$_4$N$^+$F$^-$ × 3 H$_2$O]/M	k_{obs}/s^{-1}	λ = 478 nm	k_{-1}/M^{-1} s^{-1}
5.03 × 10^{-5}	3.44 × 10^{-2}	1.05 × 10^6		9.04 × 10^6
	5.71 × 10^{-2}	1.64 × 10^6		
	1.28 × 10^{-1}	1.76 × 10^6		
	1.93 × 10^{-1}	2.70 × 10^6		
	2.27 × 10^{-1}	3.24 × 10^6		
	2.85 × 10^{-1}	3.25 × 10^6		

$k_{obs} = 9.04 \times 10^6$ [F$^-$] + 8.79 × 10^5
$R^2 = 0.9307$

[11-PPh$_3$] / M	[Bu$_4$N$^+$F$^-$ × 3 H$_2$O]/M	k_{obs}/s^{-1}	λ = 464 nm	k_{-1}/M^{-1} s^{-1}
1.39 × 10^{-4}	1.71 × 10^{-3}	3.43 × 10^7		(2.0 × 10^8)^a
	3.68 × 10^{-3}	3.48 × 10^7		
	4.92 × 10^{-3}	3.40 × 10^7		
	6.71 × 10^{-3}	3.51 × 10^7		
	8.03 × 10^{-3}	3.45 × 10^7		
	9.28 × 10^{-3}	3.58 × 10^7		
	1.14 × 10^{-2}	3.49 × 10^7		
	1.24 × 10^{-2}	3.68 × 10^7		
	1.51 × 10^{-2}	3.70 × 10^7		

$k_{obs} = 2.01 \times 10^8$ [F$^-$] + 3.36 × 10^7
$R^2 = 0.6712$

^{a} This value has to be considered approximate due to the low slope of the k_{obs} vs. [F$^-$] plot.

Table 2.16. Rate constants k_{obs} for the reaction of various benzhydrylium ions with fluoride in 10AN90W.

[15-PPh$_3$] / M	[K$^+$F$^-$]/M	k_{obs}/s^{-1}	λ = 500 nm	k_{-1}/M^{-1} s^{-1}
5.99 × 10^{-5}	1.12 × 10^{-1}	1.55 × 10^5		1.37 × 10^5
	1.91 × 10^{-1}	1.65 × 10^5		
	2.76 × 10^{-1}	1.79 × 10^5		
	4.61 × 10^{-1}	2.05 × 10^5		
	7.23 × 10^{-1}	2.38 × 10^5		

$k_{obs} = 1.37 \times 10^5$ [F$^-$] + 1.40 × 10^5
$R^2 = 0.9987$

2. Nucleofugality and Nucleophilicity of Fluoride in Protic Solvents

Table 2.16. (continued)

[14-PPh$_3$] / M	[K$^+$F$^-$]/M	k_{obs}/s^{-1}	λ = 500 nm	$k_{-1}/M^{-1}\,s^{-1}$
6.08 × 10^{-5}	9.51 × 10^{-2}	3.38 × 10^5	$k_{obs} = 3.28 \times 10^5\,[F^-] + 3.15 \times 10^5$ $R^2 = 0.9944$	3.28 × 10^5
	1.55 × 10^{-1}	3.74 × 10^5		
	2.93 × 10^{-1}	4.13 × 10^5		
	4.83 × 10^{-1}	4.72 × 10^5		
	6.98 × 10^{-1}	5.43 × 10^5		

[13-PPh$_3$] / M	[K$^+$F$^-$]/M	k_{obs}/s^{-1}	λ = 478 nm	$k_{-1}/M^{-1}\,s^{-1}$
5.73 × 10^{-5}	1.11 × 10^{-1}	1.18 × 10^6	$k_{obs} = 1.20 \times 10^6\,[F^-] + 1.04 \times 10^6$ $R^2 = 0.9993$	1.20 × 10^6
	1.24 × 10^{-1}	1.18 × 10^6		
	4.40 × 10^{-1}	1.58 × 10^6		
	6.10 × 10^{-1}	1.76 × 10^6		
	8.37 × 10^{-1}	2.05 × 10^6		

Table 2.17. Rate constants k_{obs} for the reaction of various benzhydrylium ions with fluoride in 100W.

[15-PPh$_3$] / M	[K$^+$F$^-$]/M	k_{obs}/s^{-1}	λ = 500 nm	$k_{-1}/M^{-1}\,s^{-1}$
8.34 × 10^{-5}	5.65 × 10^{-2}	1.84 × 10^5	$k_{obs} = 1.09 \times 10^5\,[F^-] + 1.79 \times 10^5$ $R^2 = 0.9893$	1.09 × 10^5
	1.82 × 10^{-1}	2.04 × 10^5		
	5.30 × 10^{-1}	2.32 × 10^5		
	9.08 × 10^{-1}	2.80 × 10^5		

2. Nucleofugality and Nucleophilicity of Fluoride in Protic Solvents

Table 2.17. (continued)

[13-PPh$_3$] / M	[K$^+$F$^-$]/M	k_{obs}/s^{-1}	λ = 478 nm	k_{-1}/M^{-1} s^{-1}
5.97 × 10^{-5}	1.12 × 10^{-1}	1.13 × 10^6		1.02 × 10^6
	1.73 × 10^{-1}	1.20 × 10^6		
	3.30 × 10^{-1}	1.36 × 10^6		
	4.29 × 10^{-1}	1.48 × 10^6	k_{obs} = 1.02 × 10^6 [F$^-$] + 1.02 × 10^6	
	5.70 × 10^{-1}	1.56 × 10^6	R^2 = 0.9948	
	8.71 × 10^{-1}	1.92 × 10^6		

The solubility of 14-PPh$_3$ in 100W was insufficient for the photogeneration of the carbocation 14$^+$.

Table 2.18. Rate constants k_{obs} for the reaction of various benzhydrylium ions with fluoride in 80E20W.

[16-PPh$_3$] / M	[Bu$_4$N$^+$F$^-$ × 3 H$_2$O]/M	k_{obs}/s^{-1}	λ = 513 nm	k_{-1}/M^{-1} s^{-1}
2.41 × 10^{-5}	2.78 × 10^{-2}	5.47 × 10^5		3.83 × 10^6
	3.02 × 10^{-2}	4.97 × 10^5		
	4.95 × 10^{-2}	6.96 × 10^5		
	5.29 × 10^{-2}	6.23 × 10^5		
	6.97 × 10^{-2}	7.33 × 10^5		
	1.06 × 10^{-1}	8.11 × 10^5	k_{obs} = 3.83 × 10^6 [F$^-$] + 4.37 × 10^5	
	1.42 × 10^{-1}	9.86 × 10^5	R^2 = 0.9387	

[15-PPh$_3$] / M	[Bu$_4$N$^+$F$^-$ × 3 H$_2$O]/M	k_{obs}/s^{-1}	λ = 500 nm	k_{-1}/M^{-1} s^{-1}
4.78 × 10^{-5}	2.44 × 10^{-2}	2.11 × 10^6		1.06 × 10^7
	5.47 × 10^{-2}	2.63 × 10^6		
	1.04 × 10^{-1}	2.98 × 10^6		
	1.38 × 10^{-1}	3.19 × 10^6		
	1.69 × 10^{-1}	3.95 × 10^6		
	1.74 × 10^{-1}	3.94 × 10^6	k_{obs} = 1.06 × 10^7 [F$^-$] + 1.93 × 10^6	
	2.16 × 10^{-1}	4.06 × 10^6	R^2 = 0.949	

2. Nucleofugality and Nucleophilicity of Fluoride in Protic Solvents

Comparison of k_{solv} and $k_{solv\ calcd.}$:

Table 2.19. Comparison of experimental rate constants for the reactions of **13^+-17^+** with the solvent (k_{solv}) and calculated rate constants $k_{solv\ calcd.}$ calculated using equation 2.3.

Ar_2CH^+	Solvent	$N / s_N{}^a$	k_{solv}/s^{-1}	$k_{solv\ calcd.}/s^{-1}$	$k_{solv}/k_{solv\ calcd.}$
17^+	90AN10W	4.56 / 0.94	7.05×10^3	1.02×10^3	6.92
16^+			2.12×10^4	5.75×10^3	3.68
15^+			6.78×10^4	1.93×10^4	3.51
17^+	100M	7.54 / 0.92	7.39×10^5	4.85×10^5	1.52
16^+			2.44×10^6	2.64×10^6	0.92
15^+			$8.4 \times 10^{6\ b}$	8.65×10^6	0.97
14^+			1.97×10^7	3.15×10^7	0.63
17^+	80AN20W	5.02 / 0.89	9.38×10^3	1.81×10^3	5.19
16^+			3.08×10^4	9.32×10^3	3.30
15^+			9.49×10^4	2.94×10^4	3.23
14^+			2.52×10^5	1.02×10^5	2.46
17^+	60AN40W		1.14×10^4	$1.97 \times 10^{3\ c}$	5.79
16^+			3.16×10^4	$1.03 \times 10^{4\ c}$	3.06
15^+			9.61×10^4	$3.30 \times 10^{4\ c}$	2.92
14^+			2.81×10^5	$1.17 \times 10^{5\ c}$	2.41
13^+			1.03×10^6	$7.08 \times 10^{5\ c}$	1.45

[a] Nucleophilicity parameters for the calculation of $k_{solv\ calcd.}$ were taken from Ref. [59] [b] Rate constants for reaction with solvent taken from Ref. [41] [c] Nucleophilicity parameters for 60AN40 are not available. The nucleophilicity parameters in 67AN33W (5.05/0.90) from Ref. [59] were used instead for the calculation of $k_{solv}/k_{calcd.}$.

2. Nucleofugality and Nucleophilicity of Fluoride in Protic Solvents

¹H-NMR and ¹³C-NMR spectra:

11-F

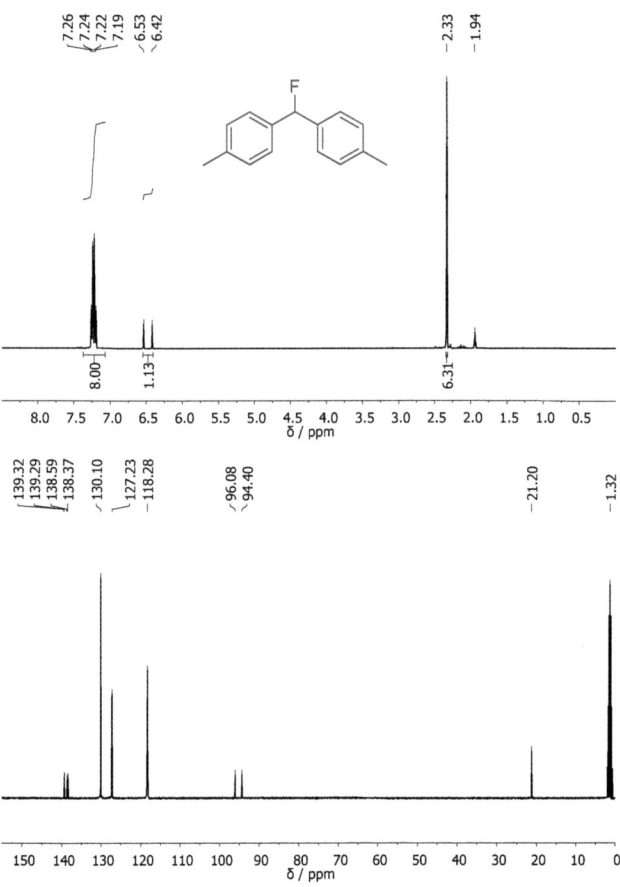

2. Nucleofugality and Nucleophilicity of Fluoride in Protic Solvents

15-F

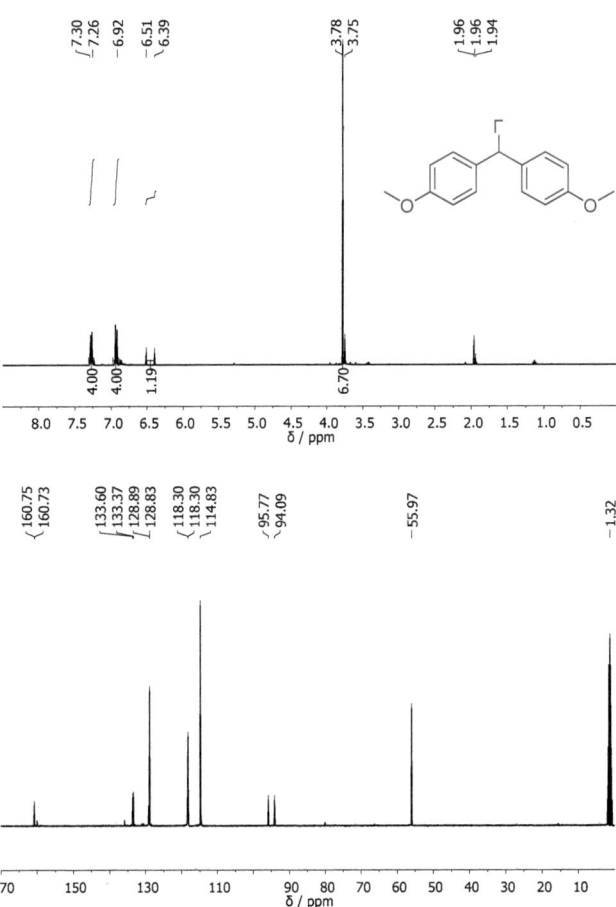

2.6. References

(1) Gribble, G. W. *J. Chem. Educ.* **2004**, *81*, 1441-1449.

(2) Gribble, G. W. *The Handbook of Environmental Chemistry Part N Organofluorines* Springer-Verlag Berlin Heidelberg, 2002.

(3) Vaillancourt, F. H.; Yeh, E.; Vosburg, D. A.; Garneau-Tsodikova, S.; Walsh, C. T. *Chem. Rev.* **2006**, *106*, 3364-3378.

(4) Ismail, F. M. D. *J. Fluorine Chem.* **2002**, *118*, 27-33.

(5) Purser, S.; Moore, P. R.; Swallow, S.; Gouverneur, V. *Chem. Soc. Rev.* **2008**, *37*, 320-330.

(6) O'Hagan, D. *J. Fluorine Chem.* **2010**, *131*, 1071-1081.

(7) Furuya, T.; Kamlet, A. S.; Ritter, T. *Nature* **2011**, *473*, 470-477.

(8) Furuya, T.; Strom, A. E.; Ritter, T. *J. Am. Chem. Soc.* **2009**, *131*, 1662-1663.

(9) Tang, P.; Furuya, T.; Ritter, T. *J. Am. Chem. Soc.* **2010**, *132*, 12150-12154.

(10) Watson, D. A., Su, M., Teverovskiy, G.; Zhang, Y., García-Fortanet, J.; Kinzel, T., Buchwald, S. L. *Science* **2009**, *325*, 1661-1664.

(11) Sachinidis, J. I.; Poniger, S.; Tochon-Danguy, H. J. *Curr. Radiopharm.* **2010**, *3*, 248-253.

(12) Schirrmacher, R.; Wangler, C.; Schirrmacher, E. *Mini-Rev. Org. Chem.* **2007**, *4*, 317-329.

(13) Kim, D. W.; Choe, Y. S.; Chi, D. Y. *Nucl. Med. Biol.* **2003**, *30*, 345-350.

(14) Elander, N.; Jones, J. R.; Lu, S.-Y.; Stone-Elander, S. *Chem. Soc. Rev.* **2000**, *29*, 239-249.

(15) J. Adams, D.; H. Clark, J. *Chem. Soc. Rev.* **1999**, *28*, 225-231.

(16) Ingold, C. K.; Ingold, E. H. *J. Chem. Soc.* **1928**, 2249-2262.

(17) Cooper, K. A.; Hughes, E. D. *J. Chem. Soc.* **1937**, 1183-1187.

(18) Swain, C. G.; Scott, C. B. *J. Am. Chem. Soc.* **1953**, *75*, 246-248.

(19) Swain, C. G.; Mosely, R. B. *J. Am. Chem. Soc.* **1955**, *77*, 3727-3731.

(20) Coverdale, A. K.; Kohnstam, G. *J. Chem. Soc.* **1960**, 3806-3731.

(21) Swain, C. G.; Knee, T. E. C.; MacLachlan, A. *J. Am. Chem. Soc.* **1960**, *82*, 6101-6104.

(22) Toteva, M. M.; Richard, J. P. *J. Am. Chem. Soc.* **2002**, *124*, 9798-9805.

(23) Kevill, D. N.; D'Souza, M. J. *J. Org. Chem.* **2004**, *69*, 7044-7050.

(24) Swain, C. G.; Mosely, R. B.; Bown, D. E. *J. Am. Chem. Soc.* **1955**, *77*, 3731-3737.

(25) Landini, D.; Maia, A.; Rampoldi, A. *J. Org. Chem.* **1989**, *54*, 328-332.

(26) de Oliveira Baptista M. J. V.; Widdowson, D. A. *J. Chem. Soc., Perkin Trans.1* **1978**, 295-298.

(27) Denegri, B.; Streiter, A.; Juric, S.; Ofial, A. R.; Kronja, O.; Mayr, H. *Chem. Eur. J.* **2006**, *12*, 1648-1656.

(28) Denegri, B.; Ofial, A. R.; Juric, S.; Streiter, A.; Kronja, O.; Mayr, H. *Chem. Eur. J.* **2006**, *12*, 1657-1666.

(29) Streidl, N.; Antipova, A.; Mayr, H. *J. Org. Chem.* **2009**, *74*, 7328-7334.

(30) Streidl, N.; Branzan, R.; Mayr, H. *Eur. J. Org. Chem.* **2010**, 4205-4210.

(31) Streidl, N.; Denegri, B.; Kronja, O.; Mayr, H. *Acc. Chem. Res.* **2010**, *43*, 1537-1549.

(32) Matic, M.; Denegri, B.; Kronja, O. *Eur. J. Org. Chem.* **2010**, 6019-6024.

(33) Juric, S.; Denegri, B.; Kronja, O. *J. Org. Chem.* **2010**, *75*, 3851-3854.

(34) Ando, T.; Cork, D. G.; Fujita, M.; Kimura, T.; Tatsuno, T. *Chem. Lett.* **1988**, 1877-1878.

(35) Minegishi, S.; Loos, R.; Kobayashi, S.; Mayr, H. *J. Am. Chem. Soc.* **2005**, *127*, 2641-2649.

(36) Ammer, J.; Baidya, M.; Kobayashi, S.; Mayr, H. *J. Phys. Org. Chem.* **2010**, *23*, 1029-1035.

(37) Mayr, H.; Bug, T.; Gotta, M. F.; Hering, N.; Irrgang, B.; Janker, B.; Kempf, B.; Loos, R.; Ofial, A. R.; Remennikov, G.; Schimmel, H. *J. Am. Chem. Soc.* **2001**, *123*, 9500-9512.

(38) Sun, H.; DiMagno, S. G. *J. Am. Chem. Soc.* **2005**, *127*, 2050-2051.

(39) For the photo-heterolysis of quaternary phosphonium salts see: a) Alonso E. O.; Johnston L.; Scaiano J. C.; Toscano V.G. *J. Am. Chem. Soc.* **1990**, *112*, 1270-1271; b) Alonso E. O.; Johnston L.; Scaiano J. C.; Toscano V.G. *Can. J. Chem.* **1992**, *70*, 1784-1794; c) Imrie C.; Modro T. A.; Rohwer E. R.; Wagener C. C. P. *J. Org.*

Chem. **1993**, *58*, 5643-5649; d) Imrie C.; Modro T. A.; Wagener C. C. P. *J. Chem. Soc., Perkin Trans. 2* **1994** 1379-1382.

(40) Minegishi, S.; Kobayashi, S.; Mayr, H. *J. Am. Chem. Soc.* **2004**, *126*, 5174-5181.

(41) McClelland, R. A.; Kanagasabapathy, V. M.; Steenken, S. *J. Am. Chem. Soc.* **1988**, *110*, 6913-6914.

(42) Pham, T. V.; McClelland, R. A. *Canadian Journal of Chemistry* **2001**, *79*, 1887-1897.

(43) Minegishi, S.; Mayr, H. *J. Am. Chem. Soc.* **2002**, *125*, 286-295.

(44) Calculated by equation 2.1 using $E_f = -3.44$, $N = 3.84$, and $s_f = 0.96$ from Ref. [31].

(45) Because of the low accuracy of this rate constant (see plot k_{obs} vs. [F⁻] in table 2.15) and the proximity of the diffusion limit, this rate constant was not included in the correlations of Figure 5.

(46) Mayr, H.; Ofial, A. R. *Angew. Chem.* **2006**, *118*, 1876- 1886; *Angew. Chem. Int. Ed.* **2006**, *45*, 1844-1854.

(47) Cowie, G. R.; Fitches, H. J. M.; Kohnstam, G. *J. Chem. Soc.* **1963**, 1585-1593.

(48) Denegri, B.; Kronja, O. *J. Phys. Org. Chem.* **2009**, *22*, 495-503.

(49) The nucleophilicity of F⁻ in 98AN2W was not included in Figure 8, because the relative reactivity of F⁻ in this solvent is strongly dependent on the nature of the nature of the electrophile due to the greatly differing sensitivity parameter s_N (Table 2.5)

(50) Spencer, J. F.; Le Pla, M. *Z. Anorg. Chem.* **1910**, *65*, 10-15.

(51) Baidya, M.; Kobayashi, S.; Mayr, H. *J. Am. Chem. Soc.* **2010**, *132*, 4796-4805.

(52) Padmanaban, M.; Biju, A. T.; Glorius, F. *Org. Lett.* **2011**, *13*, 98-101.

(53) Smith, J.; Tedder, J. M. *J. Chem. Res. Synop.* **1988**, 108-109.

(54) Shpanko, I. V.; Serebryakov, I. M.; Oleinik, N. M.; Panchenko, B. V. *Org. React. (Tartu)* **1996**, *30*, 3-6.

(55) Van, P. T.; McClelland, R. A. *Can. J. Chem.* **2001**, *79*, 1887-1897.

(56) Carlier, P. R.; Zhao, H.; MacQuarrie-Hunter, S. L.; DeGuzman, J. C.; Hsu, D. C. *J. Am. Chem. Soc.* **2006**, *128*, 15215-15220.

(57) Shigenori, O.; Masaki, S.; Zen-ichi, T.; Takeda Chemical Industries, Ltd., Japan . **2001**, p US6248766 B6248761, 6242001.

2. Nucleofugality and Nucleophilicity of Fluoride in Protic Solvents

(58)   Huisgen, R. *Angewandte Chemie* **1970**, *82*, 783-794.
(59)   Minegishi, S.; Kobayashi, S.; Mayr, H. *J. Am. Chem. Soc.* **2004**, *126*, 5174-5181.

3. Can One Predict Changes from S_N1 to S_N2 Mechanisms?

Most kinetic measurements (e.g., determination of nucleophilicity of amines in DMSO) were performed by Dr. Thanh Binh Phan. The GC product studies, the kinetic of the reaction of **8**-Br in DMSO in the presence of *n*-PrNH$_2$ and the determination of activation parameters were performed by the author of this thesis.

3.1. Introduction

Nucleophilic displacement reactions at C(sp^3) centers[1] proceed either with simultaneous breaking and forming of the involved bonds (S_N2 or A_ND_N)[2] or via a mechanism where breaking of the old bond precedes formation of the new bond (S_N1 or D_N+A_N).[1] The borderline between these two mechanisms has been the subject of considerable controversy. In contrast to Ingold who considered S_N1 and S_N2 as discrete processes,[1b] it has been suggested that a clear-cut distinction between these two mechanisms is impossible because there is a gradual transformation of an S_N2 into an S_N1 mechanism as the transition state develops more carbocation character.[3-6] Winstein's concept of different types of ion-pairs[4] was extended by Sneen who suggested that the entire S_N1-S_N2 mechanistic spectrum could be fitted into a simple scheme involving ion-pair intermediates.[5] Schleyer and Bentley criticized this concept and suggested that there is a gradation of mechanism between the S_N1 and S_N2 extremes with varying degrees of nucleophilic participation by the solvents.[6,7] The intermediates in the borderline region were considered as "nucleophilically solvated ion pairs"[6] which look like the transition states of S_N2 reactions but are energy minima not maxima. They coined the term "S_N2 intermediate" mechanism.[6] Support for the operation of concurrent S_N1 and S_N2 reactions in the borderline cases came from kinetic investigations of nucleophilic substitutions under nonsolvolytic conditions, i.e., under conditions where the concentration of the nucleophile could be varied.[8] Nucleophilic displacement reactions of benzhydryl thiocyanates with labelled *SCN$^-$ in acetonitrile and acetone,[9] of benzhydryl chlorides with labeled Cl$^-$ and Br$^-$ and of benzhydryl bromide with Br$^-$, Cl$^-$, and N$_3^-$ as well as with amines followed the rate law 3.1 with a nucleophile-independent term k_1 and a nucleophile-dependent term k_2.[10,11]

3. Can One Predict Changes from S_N1 to S_N2 Mechanisms?

$$-d[R\text{-}X]/dt = [R\text{-}X]\,(k_1 + k_2[\text{Nu}]) \quad (3.1)$$

Yoh and Fujio et al. studied the kinetics of the reactions of benzyl halides and tosylates with amines.[12,13] While acceptor-substituted benzyl derivatives reacted exclusively by the S_N2 mechanism, donor-substituted benzyl derivatives, such as p-methoxybenzyl bromide, followed the rate law of equation 3.1. This observation was considered "convincing evidence for the occurrence of simultaneous S_N1 and S_N2 mechanisms".[14a,b] Concurrent stepwise and concerted substitutions have also been reported by Amyes and Richard for the reactions of azide ions with 4-methoxybenzyl derivatives in trifluoroethanol/water mixtures.[14c] Analogous rate laws have been observed by Katritzky for alkyl and benzyl group transfers from N-alkyl and N-benzyl pyridinium ions to various nucleophiles.[15] Because of the manifold of examples which demonstrate the duality of the two mechanisms the question arises whether the change from one to the other mechanism can be predicted. Jencks and Richard based the differentiation of the mechanistic alternatives on the lifetimes of the potential intermediates.[16] It has been argued that nucleophilic aliphatic substitutions generally occur by the stepwise S_N1 mechanism when the intermediate carbocations exist in energy wells for at least the time of a bond vibration ($\approx 10^{-13}$s) and that the change to the S_N2 mechanism is "enforced" when the energy well for the intermediate disappears. Convincing support for this hypothesis has been derived from the selectivities of carbocations ($k_{\text{azide}}/k_{\text{ROH}}$) which were solvolytically generated in alcoholic solutions of ionic azides.[16c,d,17]

We have reported that the rates of reactions of carbocations with nucleophiles can be calculated by equation 3.2, where E is a carbocation-specific electrophilicity parameter, and s and N are nucleophile-specific parameters.[18-20]

$$\lg k_{20\,°C} = s_N(E+N) \quad (3.2)$$

While the confidence limit of equation 3.2 is generally a factor of 10-100 in the presently covered reactivity range of forty orders of magnitude, the predictive power of equation 3.2 is much better for reactions of benzhydrylium ions (factor 2-3), because benzhydrylium

3. Can One Predict Changes from S_N1 to S_N2 Mechanisms?

ions have been used as reference electrophiles for deriving the nucleophile-specific parameters s_N and N. Whether equation 3.2 can be used to predict the change from S_N1 to S_N2 mechanism on the basis of the lifetime hypothesis by Jencks and Richard was investigated. For that purpose, rates and products of the reactions of benzhydryl bromides (**2,6,8,9,10**)-Br with amines in DMSO were investigated, which yield benzhydryl amines (**2,6,8,9,10**)-NR$_2$, benzophenones (**2,6,8,9,10**)=O, and benzhydrols (**2,6,8,9,10**)-OH. Scheme 3.1 shows that for each of the products N°-NR$_2$, N°=O and N°-OH formation via the S_N1 process (k_1) or the S_N2 process (k_2 and k_1') has to be considered. In the following, it will be shown that the pathways k_1' and k_N can be excluded.

Scheme 3.1. Reactions of Benzhydryl Bromides with Amines in DMSO.

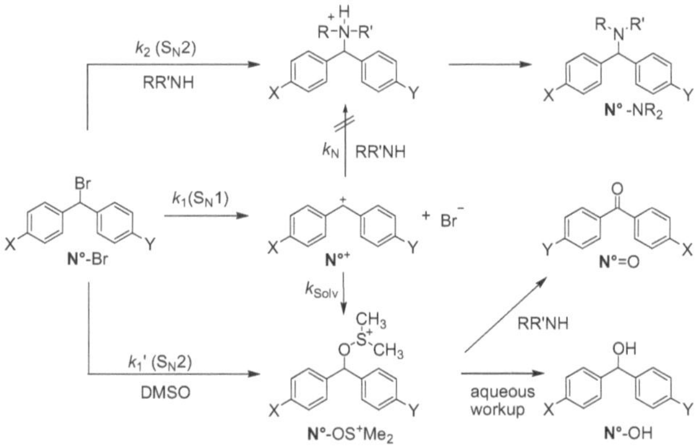

3.2. Experimental Section

Conductimetric Measurements of Nucleophilic Substitutions.

Dissolution of the benzhydryl bromides (**2,6,8,9,10**)-Br in DMSO or in solutions of amines in DMSO led to an increase of conductivity due to the generation of HBr, which reacted with excess amine to give the hydrobromide salt. The rates of these reactions were followed by conductometry (conductometers: Tacussel CD 810 or Radiometer Analytical CDM 230,

3. Can One Predict Changes from S_N1 to S_N2 Mechanisms?

Pt electrode: WTW LTA 1/NS), while the temperature of the solutions was kept constant (20.0 ± 0.1 °C) by using a circulating bath thermostat. The correlation between conductance and the concentration of liberated HBr was determined by injecting 0.25 mL portions of 0.11 M acetonitrile solutions of the rapidly ionizing benzhydryl bromide **10**-Br into 30.0 mL of a 0.34 M solution of piperidine in DMSO. After the conductivity had reached a constant value (typically 300 s), another portion of benzhydryl bromide was added. As depicted in the inset of Figure 1, the conductivity increased linearly with the concentration of released HBr, even at higher concentrations than used for the kinetic experiments.

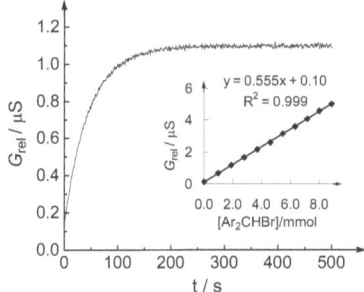

Figure 3.1. Exponential increase of conductivity during the reaction of 4-methylbenzhydryl bromide **10**-Br with 0.2 M piperidine in DMSO. Calibration in the inset: Conductivity at t_∞ is proportional to the concentration of substrate.

Photometric Measurements of the Reactions of the Benzhydrylium Tetrafluoroborates with Amines

The rates of the reactions of benzhydrylium tetrafluoroborates with amines were studied in DMSO solutions. All amines were used as free bases. As the reactions of the colored benzhydrylium ions with amines gave rise to colorless products, the reactions could be followed by employing UV-Vis spectroscopy at the absorption maxima of the benzhydrylium ions. The rates were determined by using a Hi-Tech SF-61DX2 stopped-flow spectrophotometer system (controlled by Hi-Tech KinetAsyst2 software). Amine concentrations at least 10 times higher than the benzhydrylium ion concentrations were usually employed, resulting in pseudo-first-order kinetics with an exponential decay of the

3. Can One Predict Changes from S_N1 to S_N2 Mechanisms?

concentrations of the benzhydrylium ions Ar_2CH^+ (**N°⁺**). First-order rate constants k_{obs} (s^{-1}) were obtained by least-squares fitting of the absorbance data (averaged from at least five kinetic runs at each amine concentration) to the single-exponential equation 3.3.

$$dA/dt = A_0\, e^{(-k_{obs}t)} + C \qquad (3.3)$$

Laser-Flash Photolysis.

Laser-flash photolysis methods were employed for determinating the rates of the reactions of Ar_2CH^+ (**N°⁺**) with DMSO in acetonitrile. For that purpose, benzhydrylium cations were generated by irradiation of Ar_2CHCl (**N°-Cl**) in DMSO/acetonitrile with a Continuum PL9010 Nd:YAG laser-flash apparatus (λ=266 nm; power/puls of ca. 50 mJ) in a quartz cell. The rate constants were determined by observing the time dependent decay of the UV-vis absorptions of the benzhydryl cations. The pseudo-first-order rate constants were obtained by fitting the decay of the UV-vis absorptions to the exponential function 3.3.

Product Studies.

Product studies were carried out for several representative systems to examine the ratios of the products formed during the reaction of benzhydryl bromides ((**2,6,8,9,10**)-Br) solution 0.2 M of amines in DMSO. For that purpose, 0.2 M solutions of the amine in DMSO were combined with 0.1 equivalent of benzhydryl bromide (**2,6,8,9,10**)-Br. After 24 h, the reaction mixtures were quenched with water and extracted with diethyl ether. After evaporation, the residue was diluted with acetone containing a defined amount of hexadecane (internal standard, ≈ 1 × 10^{-3} M). Aliquots of the solutions were analyzed with a Thermo Focus gas chromatograph equipped with a FID detector for the determination of the absolute concentrations. In addition, GC/MS analysis with an Agilent 6890 gas chromatograph with an Agilent 5973 mass selective detector was performed for identifying the individual peaks. For the calculation of the absolute product concentrations, the products were synthesized individually, and GC-calibrations were carried out to obtain the relative molecular response factor (RMR).

3. Can One Predict Changes from S_N1 to S_N2 Mechanisms?

3.3. Results and Discussion

Kinetics of the Nucleophilic Substitutions in DMSO:
When solutions of the benzhydryl bromides (**2,6,8,9,10**)-Br in DMSO were treated with a high excess of amines (>10 equiv), the amine concentrations remained almost constant during the reactions, and the increase of conductivity followed the exponential function 3.4, as illustrated in Figure 3.1.

$$dG/dt = G_{max}[1-e^{(-k_{obs}t)}]+C \quad (3.4)$$

Plots of k_{obs} versus the concentrations of the amines were linear (Figure 3.2) but did not go through the origin. As expressed by equation 3.5, the observed rate constants k_{obs} (see Experimental Section and supporting information of "Phan, T. B.; Nolte, C.; Kobayashi, S.; Ofial, A. R.; Mayr, H. *J. Am. Chem. Soc.* **2009**, 131, 11392-11401") can be regarded as the sum of an amine independent term k_1 and an amine-dependent term k_2[amine], which are collected in Table 3.1.

$$k_{obs} = k_1 + k_2[\text{amine}] \quad (3.5)$$

The second-order rate constants k_2 can easily be assigned to the S_N2 reactions of the amines with the benzhydryl bromides. The amine-independent term k_1, which equals the directly measured solvolysis rate constant in DMSO in the absence of a nucleophilic amine, reflects either the rate of the S_N1-type process (k_1, Scheme 3.1), the rate of an S_N2 reaction with DMSO as the nucleophile (k_1', Scheme 3.1) or a combination of both processes.

3. Can One Predict Changes from S_N1 to S_N2 Mechanisms?

Table 3.1. Rate Constants (at 20° C) for the solvolyses of the benzhydryl bromides in DMSO (s^{-1}) and for their reactions with amines in DMSO ($M^{-1}s^{-1}$).

nucleophiles	10-Br	9-Br	8-Br	6-Br	2-Br
DMSO (k_1)	6.71×10^{-3}	5.45×10^{-4}	1.36×10^{-4}	1.25×10^{-5}	2.76×10^{-6}
DABCO (k_2)	1.92×10^{-1}	5.45×10^{-2}	a	a	a
piperidine (k_2)	3.57×10^{-2}	1.69×10^{-2}	2.33×10^{-2}	9.36×10^{-3}	6.66×10^{-3}
morpholine (k_2)	2.16×10^{-2}	7.30×10^{-3}	9.51×10^{-3}	3.29×10^{-3}	2.17×10^{-3}
ethanolamine (k_2)	a	1.54×10^{-3}	2.37×10^{-3}	1.13×10^{-3}	1.25×10^{-3}
1-aminopropan-2-ol (k_2)	4.93×10^{-3}	1.45×10^{-3}	2.23×10^{-3}	8.92×10^{-4}	7.96×10^{-4}
n-PrNH$_2$ (k_2)	3.98×10^{-3}	1.33×10^{-3}	2.19×10^{-3}	1.13×10^{-3}	1.17×10^{-3}
benzylamine (k_2)	1.90×10^{-3}	6.72×10^{-4}	1.35×10^{-3}	6.30×10^{-4}	5.50×10^{-4}
diethanolamine (k_2)	a	6.37×10^{-4}	7.46×10^{-4}	2.55×10^{-4}	1.19×10^{-4}
2-amino-butan-1-ol (k_2)	b	b	3.13×10^{-4}	1.77×10^{-4}	a

a Not determined. b The k_{obs} was independent of the amine concentration see Figure 3.2

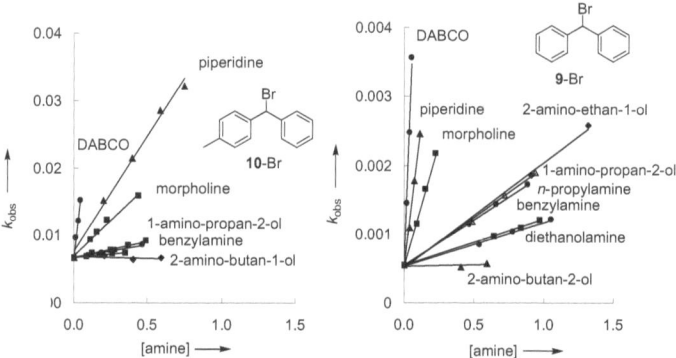

Figure 3.2. Plots of k_{obs} (s^{-1}) of the reactions of different benzhydryl bromides with amines in DMSO vs. the concentrations of the amines. (Note the different calibration of the various y-axes).

3. Can One Predict Changes from S_N1 to S_N2 Mechanisms?

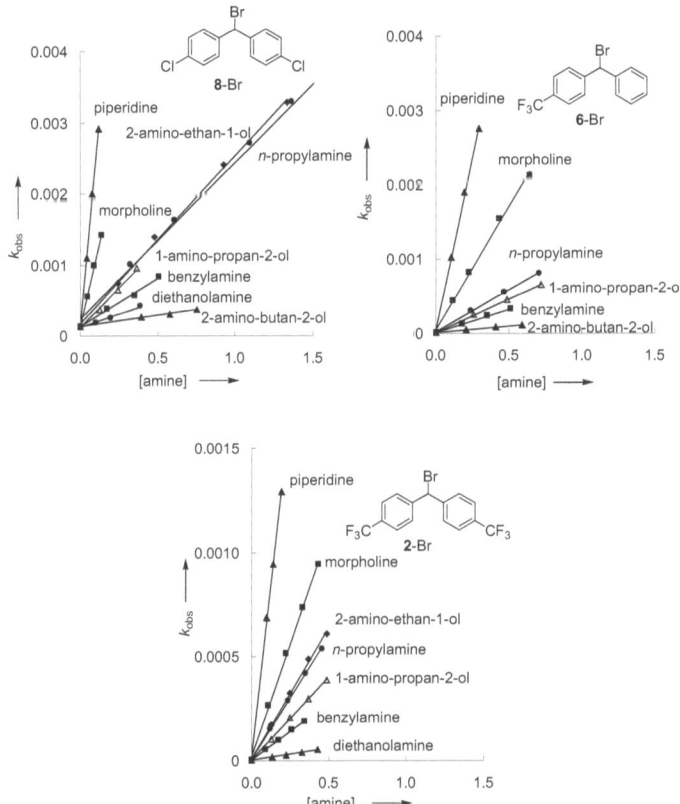

Figure 3.2. (continued)

Comparison of the rate constants k_1 in the first line of Table 3.1 with previously published solvolysis rates in alcohols[21] shows that the solvolysis rates in DMSO are comparable to those in pure ethanol but considerably smaller than in ethanol-water mixtures and 2,2,2-trifluoroethanol. Creary et al. published kinetic data for the solvolysis of adamantyl mesylates and 3-aryl-3-hydroxy-β-lactams.[22] These and our data indicate that DMSO is a solvent with a relatively high ionizing power. The horizontal lines in Table 3.1 show that

3. Can One Predict Changes from S_N1 to S_N2 Mechanisms?

variation of the substituents of the benzhydryl bromides from **10**-Br to **2**-Br affects the second order rate constants k_2 for the reactions with amines by less than a factor of 10. Accordingly, plots of lg k_2 versus $\sum\sigma$ (Figure 3.3) or any other of Hammett's substituent constants (eg. $\sigma+$)[23] (Figure 3.3) illustrate that variation of the *para*-substituents in the benzhydrylium bromides has only a marginal effect on the rate constants of the S_N2 reactions, indicating transition states $\mathbf{A}^{\#}$, where only little positive charge is developed at the benzhydryl center.

$$\left[\begin{array}{c} R\,\,\delta+\,\,\,Ar\,\,\,Ar \\ \diagdown\,\,\,\diagup \\ N\cdots\cdots C\cdots\cdots\cdot\delta-\,Br \\ R'\diagup\,\,\,\,\,\,\,\,\,\,\,\,| \\ H\,\,\,\,\,\,\,H \end{array}\right]^{\#} \quad \mathbf{A}^{\#}$$

Because of the small dependence of the second-order rate constants on the nature of the substituents, the poor correlations in Figure 3.3 are not surprising, particularly because substituent effects in diarylmethyl compounds have been reported not to be additive.[24] The poor correlation in Figure 3.3 also is in line with previous findings by Baker,[25] Jencks[26] and Bordwell[27] that nucleophilic substitutions at substituted benzyl chlorides do not follow simple Hammett correlations.

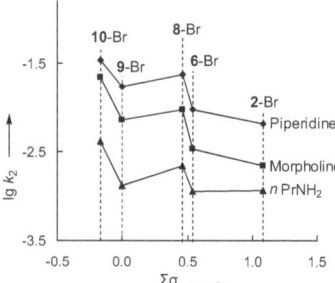

Figure 3.3. Plot of lg k_2 of the reactions of the benzhydryl bromides (**2,6,8,9,10**)-Br with amines vs. Hammett´s substituent constants σ.[16]

On the other hand, the S_N2 reactions of substituted arylethyl bromides show a continuous increase of the ρ value as the electron-donating ability of the substituents is increased, indicating a continuous change of the transition state from very tight for acceptor-

3. Can One Predict Changes from S_N1 to S_N2 Mechanisms?

substituted systems to loose transition states with more positive charge on the benzylic carbon for the S_N2 reactions of the *p*-methoxy-substituted systems.[12]

In contrast to the behavior of the second-order rate constants in Figure 3.3, the first-order rate constants k_1 (first line of Table 3.1) strongly depend on the *para*-substituents. From the plot of lg k_1 versus $\Sigma\sigma^+$, one derives a Hammett reaction constant of $\rho = -2.94$ (Figure 3.4).

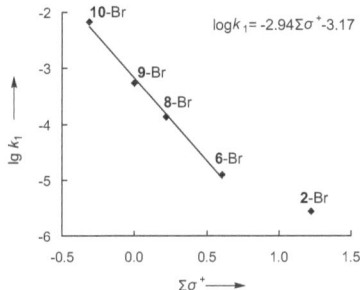

Figure 3.4. Plot of lg k_1 of the solvolysis reactions of the benzhydryl bromides (**2,6,8,9,10**)-Br in DMSO vs. Hammett's substituent constants $\sigma^{+\,23}$ (k_1 for **2**-Br not used for the correlation, see text).

The magnitude of the reaction constant ρ suggests that the amine-independent term k_1 corresponds to the ionization step of an S_N1 reaction and not to an S_N2-type attack of DMSO at the benzhydryl bromides. The bis-trifluoromethyl-substituted compound **2-Br** deviates from this correlation, however, and reacts approximately 16 times faster than extrapolated from the linear lg k_1 versus $\Sigma\sigma^+$ correlation; it will be discussed later that this deviation may be due to an S_N2-type reaction of **2-Br** with DMSO.

The 4,4′-dimethyl-substituted benzhydryl bromide (**11-Br**) reacted so fast that analogous experiments, as described in Figure 3.2, could not be performed. From the Hammett correlation
given in Figure 3.4, one can extrapolate a first-order solvolysis rate constant of 0.045 s^{-1} for the 4,4′-dimethyl-substituted benzhydryl bromide (**11-Br**).

Reaction Products

3. Can One Predict Changes from S_N1 to S_N2 Mechanisms?

As summarized in Table 3.2, the reactions of benzhydryl bromides (**2,6,8,9,10**)-Br with 0.2 M amines in DMSO give the benzhydryl amines **N°**-NR$_2$, accompanied by the benzophenones **N°**=O and the diarylmethanols **N°**-OH.

The exclusive formation of the benzhydryl amines (**2,6**)-NR$_2$ in the reaction of **2**-Br and **6**-Br with morpholine, piperidine, and n-propylamine is in line with the kinetics described in Figure 3.2: The amine-independent terms are negligible in comparison to the amine-dependent terms. Therefore, at amine concentrations of 0.2 M S_N2 reactions with amines take place almost exclusively.

Analogously, the predominant formation of amine **8**-NR$_2$ by reaction of **8**-Br with piperidine can be explained by the high S_N2 reactivity of the amine at a concentration of 0.2 M. In the reactions with the less nucleophilic morpholine, the amounts of benzhydrol **8**-OH and benzophenone **8**=O rise. Unfortunately, it was not possible to measure the product ratio obtained by the reaction of **8**-Br with n-propylamine because the GC signals of the benzhydrol and the amine overlapped. In the reactions of **9**-Br with these amines, considerable amounts of benzophenone **9**=O and benzhydrol **9**-OH were generated along with the benzhydryl amines **9**-NR$_2$, and their quantities increase with decreasing nucleophilicities of the amines.

In the reaction of the monomethyl-substituted benzhydryl bromide **10**-Br, an even larger amount of diarylmethanol **10**-OH and benzophenone **10**=O was found, while less of the benzhydryl amine **10**-NR$_2$ was formed. As with the other substrates, the yield of the amine **10**-NR$_2$ decreased in the series piperidine > morpholine > n-propylamine. While it was not possible to distinguish between diarylmethanol **10**-OH and benzophenone **10**=O by our GC analysis because both compounds had the same retention times, the GC-MS spectra showed that benzophenone **10**=O is the major products. Because the relative molar response (RMR) constant was nearly the same for the benzophenone **10**=O and the benzhydrol **10**-OH (experiment with pure compounds), their sum could be determined.

As illustrated in Scheme 3.1, the benzophenones **N°**=O as well as the benzhydrols **N°**-OH are formed through the intermediacy of the oxysulfonium ions **N°**-OS$^+$Me$_2$. In accordance with previous reports on the mechanism of the Kornblum oxidation,[28] we assume that deprotonation of the oxysulfonium ion **N°**-OS$^+$Me$_2$ at a methyl group yields a sulfur ylide, which undergoes a proton shift and cleavage of the O-S bond to yield the benzophenone

3. Can One Predict Changes from S_N1 to S_N2 Mechanisms?

$N°=O$ (Scheme 3.2). In line with this mechanism, benzhydrol $N°$-OH was not oxidized when treated with equimolar amounts of 2,6-lutidine and 2,6-lutidine hydrobromide under the conditions of the solvolysis reactions. The formation of oxysulfonium ions from alkyl halides[22] and their subsequent reactions with bases have been studied NMR spectroscopically by other groups.[29,30]

Scheme 3.2. Reaction mechanism for the generation of the benzophenone $N°=O$ from the intermediately formed oxysulfonium ion $N°$-OS⁻Me₂.

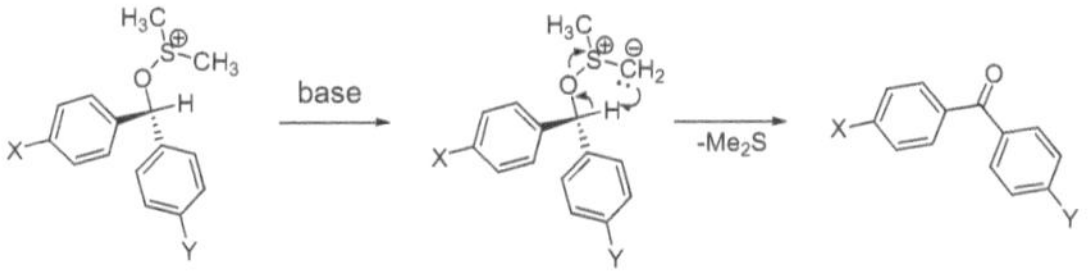

Table 3.2. Products of the reactions of the benzhydryl bromides (**2,6,8,9,10**)-Br (c = 0.02 M) with amines (10 equiv) in DMSO (20 °C).

$N°$-Br	amine	[$N°$-OH]	[$N°=O$]	[$N°$-NR₂]	[$N°$-NR₂]/ ([$N°=O$]+[$N°$-OH])
2-Br	piperidine	0	0	only	
	morpholine	0	0	only	
	n-propylamine	0	0	only	
6-Br	piperidine	0	0	only	
	morpholine	0	0	only	
	n-propylamine	0	0	only	
8-Br	piperidine	0.6	1.5	77.0	36.9
	morpholine	2.0	2.6	50.5	11.0
	n-propylamine	Alcohol cannot be separated from amine by GC			
9-Br	piperidine	0.6	7.4	69.9	8.7
	morpholine	2.7	16.1	69.8	3.7
	n-propylamine	16.6	26.7	48.4	1.1
10-Br	piperidine	35.4 [a]		41.8	1.2
	morpholine	49.8 [a]		27.1	0.5
	n-propylamine	80.5 [a]		14.6	0.2

[a] As the ketone **10**=O and the alcohol **10**-OH cannot be separated on the GC (see text), the yield refers to the sum of both compounds.

3. Can One Predict Changes from S_N1 to S_N2 Mechanisms?

Benzhydrol **9-OH** and benzophenone **9=O** were formed exclusively when benzhydryl bromide **9-Br** (0.02 M) was dissolved in a 0.2 M solution of the weakly nucleophilic 2,6-lutidine in DMSO. Figure 3.5 illustrates that the ratio [**9-OH**]/[**9=O**] obtained after aqueous workup of the solvolysis products from **9-Br** increases with reaction time. From the observation that the increase of this ratio continues after complete consumption of **9-Br**, one can derive that a precursor of **9=O** (e.g., the oxysulfonium ion **9-OS$^+$Me$_2$**) accumulates in the reaction mixture before it is slowly converted into benzophenone **9=O**.

When the mixture obtained from **9-Br** (0.02 M) and 0.2 M 2,6-lutidine in DMSO was worked up with methanol, the benzophenone **9=O** was accompanied by benzhydryl methyl ether, which may be formed by nucleophilic attack of methanol at the oxysulfonium ion **N°-OS$^+$Me$_2$**. Nucleophilic attack of impurities of water, amine, or methanol at the sulfur atom of **9-OS$^+$Me$_2$** may account for the small amount of benzhydrol **9-OH** (6.7 %) obtained under these conditions. The ratio [**9=O**]/([**9-OH**] + [**9-OMe**]) was similar to the ratio [**9=O**]/[**9-OH**] observed after aqueous workup at comparable reaction times (■ in Figure 3.5).

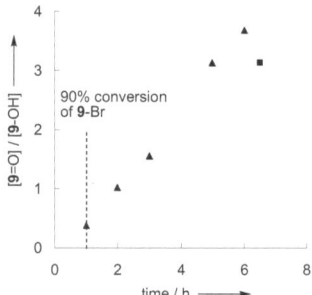

Figure 3.5. Plot of the ratios of [**9=O**]/[**9-OH**] vs. time for the reaction of benzhydryl bromide **9-Br** with 0.2 M 2,6-lutidine in DMSO after aqueous workup. Square indicates the workup with methanol (ratio equals [**9=O**]/([**9-OH**] + [**9-OMe**])).

Differentiation of S_N1 and S_N2 Processes

For each of the products **N°-NR$_2$**, **N°=O** and **N°-OH** drawn in Scheme 3.1, formation through an S_N1 (k_1) or S_N2 (k_2 and k_1') process has to be considered. With the data presented so far, it is possible to eliminate some of these reaction pathways. If the

3. Can One Predict Changes from S_N1 to S_N2 Mechanisms?

benzhydryl amines $N°$-NR_2 would be formed by an S_N1 reaction via the carbenium ions $N°^+$, which are subsequently trapped by the amines, an amine-independent rate law would result because the formation of the benzhydryl cations $N°^+$ would be rate determining. Pathway k_N of Scheme 3.1 can, therefore, be eliminated. This conclusion is confirmed by the comparison of the kinetic data with the product ratios in Table 3.3. For the reactions of different benzhydryl bromides $N°$-Br with piperidine, morpholine, and n-propylamine, the product ratios (([$N°$-NR_2]/([$N°$=O] + [$N°$-OH])), are almost equal to the ratios k_2/k_1 multiplied with the amine concentration (k_2[amine]/k_1).

Table 3.3. Comparison of rate constant ratios with the product ratios for the reactions of benzhydryl bromides $N°$-Br with 0.2 M piperidine, morpholine, and n-propylamine in DMSO at 20 °C.

	10-Br	9-Br	8-Br	6-Br
k_1/s^{-1}	6.71×10^{-3}	5.45×10^{-4}	1.36×10^{-4}	1.25×10^{-5}
reaction with piperidine				
$k_2/M^{-1} s^{-1}$	3.57×10^{-2}	1.69×10^{-2}	2.33×10^{-2}	9.36×10^{-3}
$0.2 \times k_2/k_1$	1.1	6.2	34	149.8
[$N°$-NR_2]/([$N°$=O]+[$N°$-OH])	1.2	8.7	37	only 6-NR_2
reaction with morpholine				
$k_2/M^{-1} s^{-1}$	2.16×10^{-2}	7.30×10^{-3}	9.51×10^{-3}	3.29×10^{-3}
$0.2 \times k_2/k_1$	0.64	2.7	14	263
[$N°$-NR_2]/([$N°$=O]+[$N°$-OH])	0.54	3.7	11	only 6-NR_2
reaction with n-propylamine				
$k_2/M^{-1} s^{-1}$	3.98×10^{-3}	1.33×10^{-3}	2.19×10^{-3}	1.13×10^{-3}
$0.2 \times k_2/k_1$	0.12	0.49	3.2	18
[$N°$-NR_2]/([$N°$=O]+[$N°$-OH])	0.18	1.1	-	only 6-NR_2

If the amines $N°$-NR_2 would be formed via the pathway k_N in addition to the S_N2 pathway k_2, a higher percentage of the amines $N°$-NR_2 would be expected. We will demonstrate later that the trapping of the benzhydrylium ions $N°^+$ by the solvent DMSO is so fast that the pathway k_N cannot compete with k_{solv} at amine concentrations of 0.2 M. Formal kinetics do not allow differentiation between pathways k_1 and k_1' for the formation of $N°$=O and $N°$-OH; that is, the oxysulfonium ion $N°$-OS^+Me_2 may be formed via either an S_N1 (k_1) or an S_N2 process (k_1') with the solvent DMSO. The linear Hammett plot for lg k_1 (i.e., k_{obs} in

3. Can One Predict Changes from S_N1 to S_N2 Mechanisms?

pure DMSO) with a slope of -2.94 (Figure 3.4) indicates the operation of the S_N1 pathway for most systems. If the S_N2 pathway indicated by k_1' would be operating, a similar reactivity pattern as shown in Figure 3.3 would be expected for the different benzhydryl bromides. The significant deviation of **2-Br** from the linear Hammett correlation in Figure 3.4 may be indicative of an S_N2 participation in the reaction of this acceptor-substituted benzhydryl bromide with DMSO (k_1', nucleophilic solvent participation). Further support for this interpretation will be given below.

Temperature Effect on Rate Constants

When the kinetics of the reaction of **8-Br** with morpholine in DMSO were studied at variable temperature and evaluated as described above, the rate constants summarized in Table 3.4 were obtained. Raising the temperature from 20 to 50 °C increased the first-order rate constant k_1 by a factor of 11, while the second-order rate constant k_2 increased only by a factor of 4. In accordance with the previous discussion, the stronger increase of k_1 compared with k_2 resulted in a decrease of the yield of the amine $N°$-NR_2 (Table 3.4). The ratios of the products [$N°$-NR_2]/([$N°$=O] + [$N°$-OH]) and the ratios of the rate constants k_2[amine]/k_1 again agreed within experimental error (Table 3.4), indicating that also, at elevated temperatures, amines $N°$-NR_2 are produced through the S_N2 pathway (k_2) while $N°$=O and $N°$-OH are formed via the S_N1 route (k_1).

Table 3.4. Comparison of the rate constants and product ratios for the reaction of benzhydryl bromide **8-Br** (0.02 M) with morpholine (0.2 M) at different Temperatures in DMSO.

T/°C	20	35	50
k_1/s^{-1}	1.36×10^{-4}	3.85×10^{-4}	1.45×10^{-3}
$k_2/M^{-1}s^{-1}$	9.51×10^{-3}	2.10×10^{-2}	4.20×10^{-2}
[**8**-NR_2]/M	1.01×10^{-2}	1.07×10^{-2}	8.52×10^{-3}
[**8**=O]/M	5.26×10^{-4}	1.27×10^{-3}	1.83×10^{-3}
[**8**-OH]/M	3.95×10^{-4}	2.26×10^{-4}	2.33×10^{-4}
[**8**-NR_2]/([**8**=O] + [**8**-OH])	11	7.2	4.1
$0.2k_2/k_1$	14	11	5.8

Eyring and Arrhenius plots of high quality ($R^2 = 0.9998$, see experimental part) were obtained for the second-order rate constants k_2, from which the activation parameters listed in Table 3.5 were obtained. The highly negative activation entropy (-159 J mol^{-1} K^{-1}) is in

3. Can One Predict Changes from S_N1 to S_N2 Mechanisms?

agreement with previous reports on alkylations of amines.[31] Because of the small contribution of the first-order term k_1 to the overall rate constant, the rate constants k_1 are less precise, and the resulting Eyring and Arrhenius plots are of lower quality (R^2 = 0.990). The calculated activation entropy (−117 J mol^{-1} K^{-1}) is slightly more negative than typically observed for S_N1 reactions in alcoholic and aqueous solutions.[32]

Table 3.5. Eyring and Arrhenius activation parameters for the reaction of the benzhydryl bromide **8-Br** with morpholine in DMSO.

	for k_1	for k_2
$\Delta H^{\ddagger}$/kJ mol^{-1}	59.4 ± 6.1	36.5 ± 0.6
$\Delta S^{\ddagger}$/J mol^{-1} K^{-1}	−116.7 ± 19.8	−159.0 ± 1.8
E_a/kJ mol^{-1}	62.0 ± 6.1	39.0 ± 0.5
lg A	7.2 ± 1.0	5.0 ± 0.1

Nucleophilicity Parameters N and s_N for Amines in DMSO

While N and s_N parameters for numerous amines have previously been determined in aqueous[33], acetonitrile[34] and in methanolic solution,[35] only few amines have so far been characterized in DMSO.[36] Because amine nucleophilicities in DMSO will be needed for the mechanistic analysis below, we have now determined N and s_N values for the amines which were used in this investigation in DMSO. For that purpose, the rates of reactions of amino-substituted benzhydrylium ions with amines in DMSO (Scheme 3.3) were measured under pseudo-first-order conditions (excess of amine) using the photometric method described previously.[18-20] Comparison of the rate constants listed in Table 6 with those reported in acetonitrile[34] and water[33] shows that the amines react roughly 4-10 times faster in acetonitrile and 100 times faster in DMSO than in water due to the weaker solvation in the nonprotic solvents.

3. Can One Predict Changes from S_N1 to S_N2 Mechanisms?

Scheme 3.3. Reactions of amines with benzhydrylium tetrafluoroborates in DMSO.

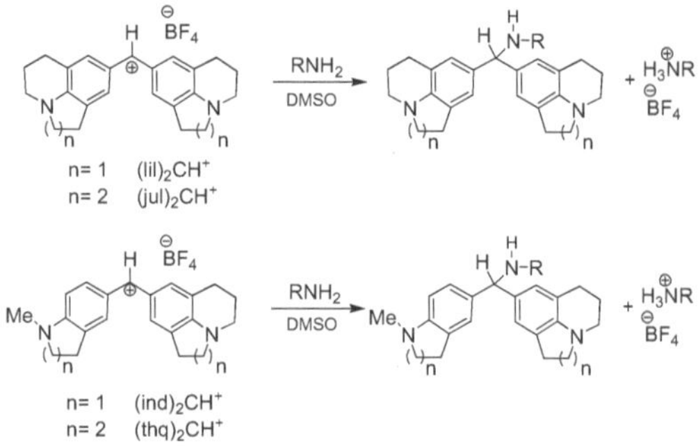

n= 1 (lil)$_2$CH$^+$
n= 2 (jul)$_2$CH$^+$

n= 1 (ind)$_2$CH$^+$
n= 2 (thq)$_2$CH$^+$

Plots (Figure 3.6) of the second-order rate constants given in Table 3.6 versus the electrophilicity parameters E of the benzhydrylium ions were linear as required by equation 3.2, and yielded the N and s_N parameters for amines in DMSO which are given in Table 3.6.

3. Can One Predict Changes from S_N1 to S_N2 Mechanisms?

Table 3.6. Second-order rate constants for the reactions of amino-substituted benzhydrylium ions with amines in DMSO at 20 °C.

Amine, N (s)	Ar_2CH^+	$k_N/M^{-1}\,s^{-1}$
2-Amino-butan-1-ol	$(ind)_2CH^+$	6.08×10^3
14.39 (0.67)	$(jul)_2CH^+$	2.23×10^3
	$(lil)_2CH^+$	8.33×10^2
Benzylamine	$(thq)_2CH^+$	3.91×10^4
15.28 (0.63)	$(jul)_2CH^+$	6.60×10^3
	$(lil)_2CH^+$	2.51×10^3
1-Amino-propan-2-ol	$(ind)_2CH^+$	2.27×10^4
15.47 (0.65)	$(jul)_2CH^+$	9.31×10^3
	$(lil)_2CH^+$	3.29×10^3
Diethanolamine	$(ind)_2CH^+$	4.83×10^4
15.51 (0.70)	$(jul)_2CH^+$	1.74×10^3
	$(lil)_2CH^+$	6.19×10^3
2-Amino-ethanol	$(ind)_2CH^+$	2.87×10^4
16.07 (0.61)	$(jul)_2CH^+$	1.19×10^4
	$(lil)_2CH^+$	4.71×10^3

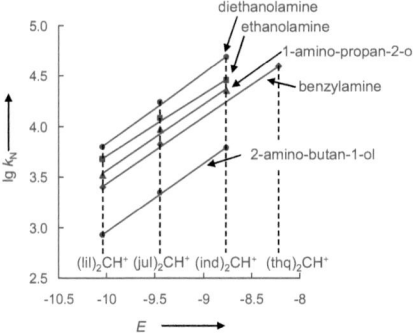

Figure 3.6. Plot of the rate constants k_N for the reactions of amines with benzhydrylium ions (DMSO, 20 °C) vs. their electrophilicity parameters E ($E = -10.04$ for $(lil)_2CH^+$, -9.45 for $(jul)_2CH^+$, -8.76 for $(ind)_2CH^+$, and -8.22 for $(thq)_2CH^+$; from ref [14]).

3. Can One Predict Changes from S_N1 to S_N2 Mechanisms?

Solvent Nucleophilicity of DMSO.
Dimethyl sulfoxide may react with electrophiles either at sulfur or at oxygen.[37] The formation of benzophenones and benzhydrols reported above indicates that the benzhydrylium ions (N^{o+}) employed in this work react at oxygen to yield the oxysulfonium ions $N^{o}\text{-OS}^{+}Me_2$ (Scheme 3.4).

Scheme 3.4. Laser-flash photolytic generation of benzhydrylium ions in MeCN/DMSO mixtures.

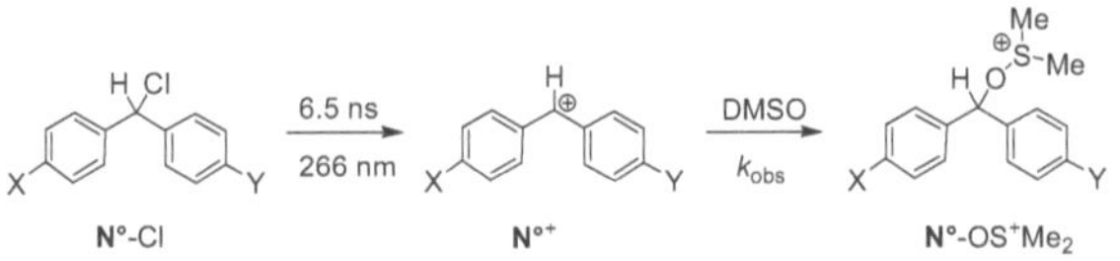

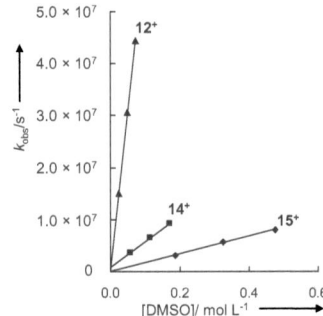

The rates of these reactions were determined by laser-flash photolysis of solutions of benzhydryl chlorides ($N^{o}\text{-Cl}$) in MeCN/DMSO mixtures and UV-vis spectrometric monitoring of the decay of the resulting benzhydrylium ions in the presence of variable concentrations of DMSO. Plots of the observed rate constants versus the concentrations of DMSO (Figure 3.7) give rise to the second-order rate constants of the reactions (Table 3.7).

Figure 3.7. Plot of k_{obs} of the reactions of the benzhydrylium ions N^{o+} with DMSO in MeCN vs. [DMSO].

3. Can One Predict Changes from S_N1 to S_N2 Mechanisms?

Table 3.7. Second-order rate constants for the reactions of benzhydrylium ions (N^{o+}) with DMSO (O-Attack) in acetonitrile.

N^{o+}	X	Y	E^a	k_2 /$M^{-1}s^{-1}$
15^+	OMe	OMe	0.00	1.69×10^7
14^+	OMe	OPh	0.61	5.00×10^7
12^+	OMe	H	2.11	6.13×10^8
$(PhO,Me)^{+\,b}$	OPh	Me	2.16	7.04×10^8
$(PhO,H)^{+\,b}$	OPh	H	2.90	8.60×10^8
11^+	Me	Me	3.63	2.63×10^9
10^+	Me	H	4.59	3.50×10^9
$(F,H)^{+\,b}$	F	H	5.60	4.78×10^9
9^+	H	H	5.90	3.34×10^9
8^+	Cl	Cl	6.02	4.79×10^9

^{a} Empirical electrophilicity parameter from Ref. [18]

^{b} (PhO,Me)$^+$, (PhO,H)$^+$, (F,H)$^+$

Figure 3.8 shows that the rate constants (Table 3.7) for the reactions of DMSO with benzhydrylium ions increase with the electrophilicity parameters of the benzhydrylium ions and become diffusion-controlled at $E > 4$. For that reason, all benzhydrylium ions N^{o+} generated in DMSO from benzhydryl bromides (**2,6,8,9,10**)-Br of Table 3.1 are immediately trapped by the solvent DMSO.

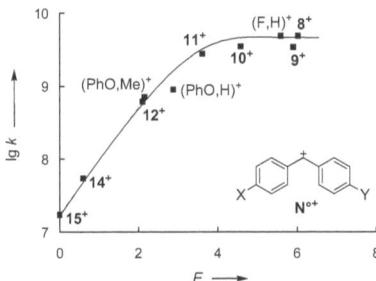

Figure 3.8. Plot of lg k (second-order rate constants/M^{-1} s^{-1}) for the reactions of DMSO with the benzhydrylium ions N^{o+} in MeCN at 20 °C vs. their electrophilicity parameters E.

3. Can One Predict Changes from S_N1 to S_N2 Mechanisms?

It can thus be explained that trapping of $(2,6,8,9,10)^+$ by amines (k_N, Scheme 1) does not occur despite the higher nucleophilicity of the amines. From the linear left part of Figure 3.8 ($k_2 < 8 \times 10^8$ M^{-1} s^{-1}), one can derive the nucleophilicity parameters $N = 9.75$ and $s_N = 0.74$ for the O-reactivity of DMSO, showing that DMSO is considerably more nucleophilic than water and ordinary alcohols.[38,39] With the assumption that the change of solvent polarity in MeCN/DMSO mixtures of different compositions does not affect the rate constants significantly, one can multiply the rate constants in Table 3.7 with 14.1 M, that is, the concentration of DMSO in 100 % DMSO to obtain the first-order rate constants of the decay in 100 % DMSO. From the plot of the first-order rate constants versus E, one derives the solvent nucleophilicity $N_1 = 11.3$ for DMSO ($N_1 = N + (\lg 14.1)/s_N$).[40] Measurements of the nucleophilic reactivity of DMSO in neat DMSO are not possible with our equipment because the laser radiation at 266 nm, which is needed for the photoionization of the benzhydryl chlorides, is absorbed by DMSO.

Calculation of Hypothetical Lifetimes of Benzhydrylium Ions in DMSO Solution in the Presence of Amines

The N and s_N parameters of the amines (Table 3.6 and from previously published data[33,36,41]) and N_1 of DMSO (Figure 3.8) can now be combined with the electrophilicity parameters E of the benzhydrylium ions to calculate rate constants for the reactions of benzhydrylium ions with these nucleophiles by equation 3.2. Many of the resulting rate constants exceed the diffusion limit. In these cases, the values $1/k$ (s), which are listed in Table 3.8, have to be considered as hypothetical lifetimes.

3. Can One Predict Changes from S_N1 to S_N2 Mechanisms?

Table 3.8. Calculated lifetimes τ ($1/k$ /s) of benzhydrylium ions in DMSO and in 1 M Solutions of various amines in DMSO.

	10⁺	**9⁺**	**8⁺**	**6⁺**	**2⁺**
Nucleophiles (N/s)	($E = 4.50$)[a]	($E = 5.60$)[a]	($E = 5.59$)[a]	($E = 6.81$)[a]	($E = 7.92$)[a]
DMSO (N_1 = 11.30/0.74)[b]	2×10^{-12}	3×10^{-13}	3×10^{-13}	4×10^{-14}	6×10^{-15}
2-amino-butan-1-ol (14.39/0.67)[c]	2×10^{-13}	4×10^{-14}	4×10^{-14}	6×10^{-15}	1×10^{-15}
benzylamine (15.59/0.63)[c]	1×10^{-13}	3×10^{-14}	3×10^{-14}	4×10^{-15}	8×10^{-16}
1-amino-propan-2-ol (15.47/0.65)[c]	1×10^{-13}	2×10^{-14}	2×10^{-14}	3×10^{-15}	6×10^{-16}
diethanolamine (15.51/0.70)[c]	1×10^{-14}	2×10^{-15}	2×10^{-15}	2×10^{-16}	4×10^{-17}
n-propylamine (15.70/0.64)[d]	1×10^{-13}	2×10^{-14}	2×10^{-14}	4×10^{-15}	8×10^{-16}
ethanolamine (16.07/0.61)[c]	3×10^{-13}	6×10^{-14}	6×10^{-14}	1×10^{-14}	2×10^{-15}
morpholine (16.96/0.67)[d]	4×10^{-15}	8×10^{-16}	8×10^{-16}	1×10^{-16}	2×10^{-17}
piperidine (17.19/0.71)[d]	3×10^{-16}	7×10^{-17}	7×10^{-17}	9×10^{-18}	1×10^{-18}
DABCO (18.80/0.70)[e]	4×10^{-17}	8×10^{-18}	8×10^{-18}	1×10^{-18}	2×10^{-19}

[a] E values taken from unpublished work by J. Ammer and C. Nolte [b] N_1 from calculation of first-order rate constant with DMSO in DMSO (see text). [c] From Table 3.6 [d] N and s_N parameters were taken from Ref. [36] N and s_N parameters in acetonitrile from Ref. [41].

The upper diagram of Figure 3.2 shows that, at a concentration of [morpholine] = 0.3 M, the observed pseudo-first-order rate constant is two times the magnitude of the intercept ($k_{obs} \approx 2k_1$); that is, at this concentration, the reaction of the methylsubstituted benzhydryl bromide (**10-Br**) with morpholine follows the S_N1 and the S_N2 mechanisms to equal extent. The calculated lifetime of 4×10^{-15} s for the reaction of the benzhydrylium ion **10⁺** with morpholine (Table 3.8) is shorter than a bond vibration ($\approx 10^{-13}$ s). According to Jencks and Richard, this relationship implies that the S_N2 mechanism will be enforced; that is, the benzhydrylium ion **10⁺** cannot exist in an encounter complex with morpholine. From the relationship k_2 (morpholine) $\approx 3k_1$ (Table 3.1), one can derive that nucleophilic assistance for breaking the C–Br bond ($\rightarrow S_N2$) is very weak and ionization (k_1) may also occur in the absence of a morpholine molecule. Only at morpholine concentrations > 0.3 M the S_N2 process will override the S_N1 process. If ionization occurs in the absence of a morpholine molecule (S_N1), the intermediate p-methylsubstituted benzhydrylium ion (**10⁺**) is rapidly trapped by the solvent DMSO (lifetime $\approx 2 \times 10^{-12}$ s), and the diffusion-controlled reaction with morpholine cannot compete.

3. Can One Predict Changes from S_N1 to S_N2 Mechanisms?

Piperidine ($k_2 \approx 5k_1$) and DABCO ($k_2 \approx 28k_1$) are stronger nucleophiles and, therefore, provide a stronger nucleophilic assistance for breaking the C-Br bond of **10-Br**. As shown in Figure 3.2, now the S_N2 process overrides the S_N1 process already at low amine concentrations, and the calculated lifetimes of 4×10^{-16} and 5×10^{-17} s are in line with Jencks' enforced concerted mechanism. Lifetimes $\tau > 10^{-14}$ s are calculated for the *p*-methylbenzhydrylium ion **10⁺** in 1 M solutions of the other amines, and Figure 3.2 shows that, in the reactions with benzylamine, 1-aminopropan-2-ol, and *n*-propylamine, the S_N1 mechanism generally dominates.

For the unsubstituted benzhydrylium ion **9⁺**, 9-times shorter lifetimes are calculated; as a consequence, the S_N2 reactions gain more weight. Morpholine ($k_2 \approx 13k_1$), piperidine ($k_2 \approx 31k_1$), and DABCO ($k_2 \approx 100k_1$) prefer the S_N2 mechanism already at low amine concentrations (> 0.08-0.01 M), in accord with calculated lifetimes of $\tau < 10^{-15}$ s. No S_N2 contribution was found for the reaction of **9-Br** with 2-aminobutan-1-ol ($\tau = 4 \times 10^{-14}$ s). For the reactions of **9-Br** with diethanolamine ($\tau = 2 \times 10^{-15}$ s), ethanolamine ($\tau = 6 \times 10^{-14}$ s), benzylamine ($\tau = 3 \times 10^{-14}$ s), 1-aminopropan-2-ol ($\tau = 2 \times 10^{-14}$ s), and n-propylamine ($\tau = 2 \times 10^{-14}$ s), lifetimes similar to the vibrational limit are calculated, and the S_N2 reactions overrated the S_N1 process only at high amine concentrations as the lifetimes were calculated for molar solutions with an amine concentration of 1 M.

Despite calculated lifetimes for the dichloro-substituted benzhydrylium ion **8⁺** which closely resemble those of the parent compound **9⁺**, Figure 3.2 shows that almost all amines prefer the S_N2 process at concentrations > 0.2 M. Only 2-aminobutan-1-ol ($\tau = 4 \times 10^{-14}$ s) allows the S_N1 mechanism to dominate at amine concentrations < 0.4 M. In agreement with calculated lifetimes $\tau < 5 \times 10^{-15}$ s, all reactions of amines with the CF₃-substituted benzhydryl bromides **6-Br** and **2-Br** studied in this work preferentially follow the S_N2 process, and the intercepts of the correlations in Figure 3.2 are negligible compared with the pseudo-first-order rate constants k_{obs} in the presence of amines. The very short lifetimes estimated for **6⁺** and **2⁺** in DMSO suggest that the first-order rate constants for the solvolyses of **6-Br** and **2-Br** in DMSO may not be due to S_N1 reactions with formation of the carbocations **6⁺** and **2⁺** because the direct nucleophilic attack of DMSO at these benzhydryl bromides should be enforced. The positive deviation of **2-Br** from the correlation of lg k_1 versus $\Sigma\sigma^+$ in Figure 3.4 is in line with a significant nucleophilic solvent

3. Can One Predict Changes from S_N1 to S_N2 Mechanisms?

participation by DMSO (k_1', S_N2). The fact that the first-order rate constant for **6-Br** matches the correlation with σ^+ in Figure 3.4 implies that in this case nucleophilic solvent participation by DMSO cannot be large.

3.4. Conclusion

In their seminal 1984 paper,[16d] Richard and Jencks concluded that a reaction can proceed concurrently through stepwise, monomolecular and concerted, bimolecular reaction mechanisms
when the intermediate has a long lifetime in the solvent, but no lifetime when it is in contact with an added nucleophilic agent. This situation has now been found when benzhydryl bromides **N°-Br** were treated with amines in DMSO. In several cases, first-order rate constants k_1 (s^{-1}) for the formation of the carbocations are of similar magnitude as the second-order rate constants k_2 (M^{-1} s^{-1}) for the concerted S_N2 reactions of the benzhydryl bromides with amines. For that reason, carbocations with short lifetimes were generated when amine molecules were not present in the vicinity, while in the same solution, concerted S_N2 reactions were enforced when amine molecules were present. The relationship ([amine]k_2)/k_1 = [**N°-NR$_2$**]/((([**N°=O**] +[**N°-OH**])) implies that the benzhydryl amines Ar$_2$CHNRR' are formed exclusively through the S_N2 process and not through trapping of the intermediate carbocations by amines. As calculated from the nucleophilicity parameters N_1 and s_N of DMSO, the intermediate benzhydrylium ions Ar$_2$CH$^+$ (**N°$^+$**), formed by the S_N1 process, are quantitatively trapped by DMSO to give the benzhydryloxysulfonium ions **N°-OS$^+$Me$_2$**, the precursors of the benzhydryl alcohols **N°-OH** and the benzophenones **N°=O**. Because the change from S_N1 to S_N2 mechanisms was observed when the lifetimes of the carbocations in the presence of amines (1 M) were calculated to be approximately 10^{-14} s by equation (3.2), the E, N, and s_N parameters proved to be suitable for predicting the preferred mechanism of the nucleophilic substitutions of benzhydryl bromides. So far, our analysis, based on Jencks' lifetime criterion, did not include the role of the leaving groups. One can expect, however, that the S_N1/S_N2 ratio for a certain substrate R-X will also depend on the leaving group and will increase with

3. Can One Predict Changes from S_N1 to S_N2 Mechanisms?

increasing nucleofugality of X. Further investigations will be presented in the following chapter.

3.5. Experimental Section, Practical Part

Only experiments conducted by the author of this thesis will be presented in this section. For the other experiments refer to the supporting information of Ref. [42]

Reactions of Benzhydryl Bromides with Amines in DMSO
Method
The reactions of the benzhydryl bromides (N°-Br) with amines in DMSO were followed by conductometry (conductometers: Radiometer Analytical CDM 230 or Tacussel CD 810, Pt electrode: WTW LTA 1/NS). The temperature of the solutions during all kinetic studies was kept constant (± 0.1 °C) by using a circulating bath thermostats.
A calibration was performed to test the dependance of the conductance on the concentration of liberated HBr. For this purpose, 0.25 mL of a 0.11 M solution of the rapidly reacting benzhydryl bromide **10-Br** in acetonitrile was injected to a 0.34 M solution of piperidine in DMSO. After the conductivity had increased to a constant value another portion of the benzhydryl bromide **10-Br** was added. As shown in Figure 3.9, the conductivity depended linearly on the concentration of added **10-Br**, i. e., released HBr.

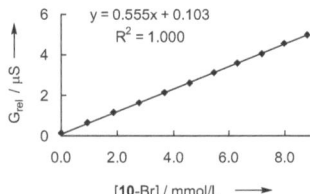

Figure 3.9. Initial concentration of the benzhydryl bromide [**10-Br**] vs. conductance at t_∞. After the addition of a portion of **10-Br**, the next conductivity value was taken when the conductivity remained constant for at least 300 s.

The first-order rate constants k_{obs} (s^{-1}) were obtained by least squares fitting of the increasing conductance to a single-exponential equation (3.4)

3. Can One Predict Changes from S_N1 to S_N2 Mechanisms?

$$dG_{rel}/dt = G_{max}[1-e^{(-k_{obs}t)}] + C \qquad (3.4)$$

Second-order rate constants k_2 (M^{-1} s^{-1}) were obtained from the slopes of linear plots of k_{obs} vs. the concentrations of the amines [Nu].

4,4'-Dichlorobenzhydryl bromide (8-Br)
20 °C, in DMSO, conductometry

Table 3.9. Individual rate constants for the reaction [8-Br] in DMSO in the presence of n-propylamine.

Nu	[8-Br]$_0$/M	[Nu]/M	k_{obs}/s^{-1}	k_2/M^{-1} s^{-1}
n-propylamine	3.42 × 10^{-4}	0.00	1.36 × 10^{-4}	**2.19 × 10^{-3}**
	5.93 × 10^{-5}	3.20 × 10^{-1}	1.01 × 10^{-3}	
	5.51 × 10^{-5}	6.10 × 10^{-1}	1.63 × 10^{-3}	
	7.21 × 10^{-5}	1.09	2.72 × 10^{-3}	
	7.18 × 10^{-5}	1.36	3.30 × 10^{-3}	
	7.10 × 10^{-5}	1.52	3.42 × 10^{-3}	

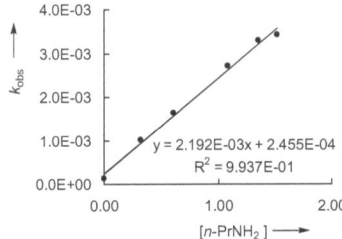

Figure 3.9. k_{obs} vs. [n-propylamine] correlations for the reactions of n-propylamine with **8-Br** in DMSO.

Temperature Dependent Kinetics of the Reaction of 4,4'-Dichlorobenzhydryl Bromide (8-Br) with Morpholine (in DMSO, conductometry)

Kinetics

For the measurements with [morpholine] = 0 mol L^{-1} an excess of at least 5 equiv 2,6-lutidine was added in order to quench released HBr and avoid autocatalysis. The rates of

3. Can One Predict Changes from S_N1 to S_N2 Mechanisms?

the investigated solvolysis reactions wererefound to be independent of the concentration of 2,6-lutidine. For the measurements at 50 °C the k_1 is the intercept of the plot [morpholine] vs. k_{obs}.

Table 3.10. Individual rate constants for the reaction [**8-Br**] in DMSO in the presence of morpholine at various temperatures.

$T/°C$	$[\text{8-Br}]_o/M$	[morpholine]/M	k_{obs}/s^{-1}	$k_2/M^{-1}\,s^{-1}$
20	3.42×10^{-4}	0.00	$1.36 \times 10^{-4\,a}$	9.50×10^{-3}
	4.85×10^{-4}	4.60×10^{-2}	$5.60 \times 10^{-4\,a}$	
	4.81×10^{-4}	9.03×10^{-2}	$1.00 \times 10^{-3\,a}$	
	4.86×10^{-4}	1.36×10^{-1}	$1.42 \times 10^{-3\,a}$	
35	2.76×10^{-4}	0.00	3.02×10^{-4}	2.10×10^{-2}
	2.70×10^{-4}	1.65×10^{-1}	3.69×10^{-3}	
	2.90×10^{-4}	3.46×10^{-2}	1.18×10^{-3}	
	2.85×10^{-4}	8.03×10^{-2}	2.18×10^{-3}	
	2.84×10^{-4}	2.16×10^{-1}	5.02×10^{-3}	
50	2.83×10^{-4}	5.98×10^{-3}	1.60×10^{-3}	4.20×10^{-2}
	2.67×10^{-4}	1.06×10^{-2}	1.87×10^{-3}	
	2.74×10^{-4}	1.51×10^{-2}	2.22×10^{-3}	
	2.73×10^{-4}	5.47×10^{-2}	1.25×10^{-3}	
	2.89×10^{-4}	6.37×10^{-2}	4.26×10^{-3}	
	2.88×10^{-4}	1.18×10^{-1}	6.56×10^{-3}	
	2.49×10^{-4}	3.58×10^{-1}	1.64×10^{-2}	

^{a} taken from the data from Ref. [42]

3. Can One Predict Changes from S_N1 to S_N2 Mechanisms?

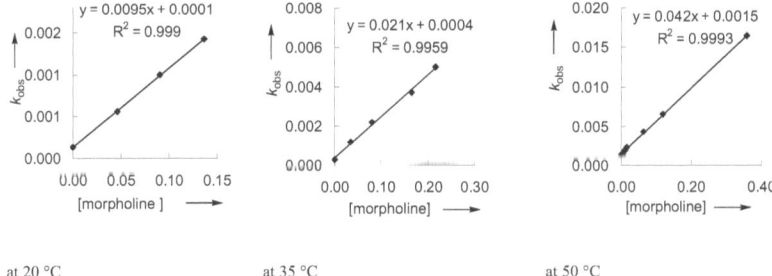

at 20 °C   at 35 °C   at 50 °C

Figure 3.10. k_{obs} vs. [morpholine] correlations for the reactions of morpholine with **8-Br** in DMSO at various temperatures.

Determination of the Eyring and Arrhenius activation parameters

Eyring and Arrhenius plots for the rate constants k_1 and k_2 of the solvolysis of 4,4'-dichlorobenzhydryl bromide in DMSO (k_1) and the reaction with morpholine (k_2); both rate constants were derived from the linear fit of k_{obs} vs. [morpholine].

Table 3.11. First- and second-order rate constants for the reaction of **8-Br** in DMSO with morpholine at various temperatures.

T/ C	k_1/s^{-1}	$k_2/M^{-1}s^{-1}$
20	1.36×10^{-4}	9.50×10^{-3}
35	3.85×10^{-4}	2.10×10^{-2}
50	1.45×10^{-3}	4.20×10^{-2}

The Eyring and Arrhenius parameters were determined by plotting of ln k vs. 1/T (T in K) resulting in the activation energies E_a as slope/R and lg A as intercepts × lg e. Plots of ln (k/T) vs. 1/T (T in K) yielded the activation enthalpies $\Delta H^{\ddagger}$ as −slope/R and activation entropies $\Delta S^{\ddagger}$ as (intercept − ln(k_B/h))/R.

3. Can One Predict Changes from S_N1 to S_N2 Mechanisms?

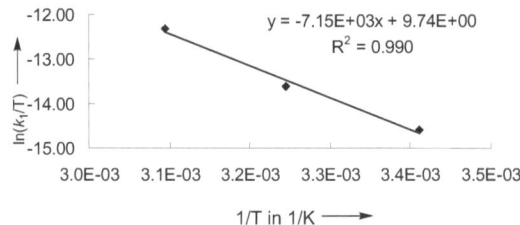

Figure 3.10. Eyring plot for k_1.

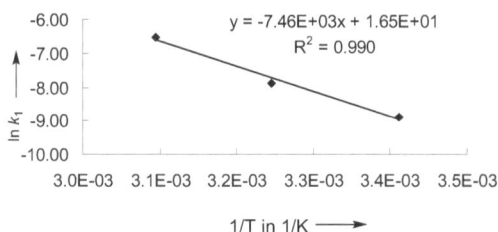

Figure 3.11. Arrhenius plot for k_1.

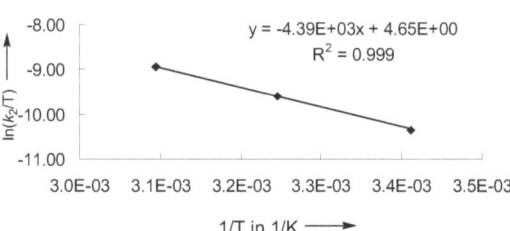

Figure 3.12. Eyring plot for k_1.

3. Can One Predict Changes from S_N1 to S_N2 Mechanisms?

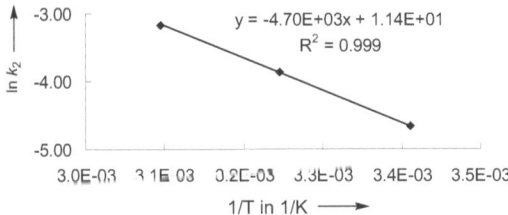

Figure 3.13. Arrhenius plot for k_2.

Table 3.12. Eyring and Arrhenius activation parameters for the rate constants k_1 and k_2 of the solvolysis of 4,4'-dichlorobenzhydryl bromide (**8-Br**) in DMSO (k_1) and the reaction with morpholine (k_2).

	for k_1	for k_2
$\Delta H^\ddagger$/kJ mol^{-1}	59.4 ± 6.1	36.5 ± 0.6
$\Delta S^\ddagger$/J mol^{-1} K^{-1}	−116.7 ± 19.8	−159.0 ± 1.8
E_a/kJ mol^{-1}	62.0 ± 6.1	39.0 ± 0.5
lg A	7.2 ± 1.0	5.0 ± 0.1

Product study

Synthesis of Compounds for GC Calibrations

General Procedure A (GP A): A solution of the amine (20 equiv) in DMSO (10 mL) was mixed with the benzhydryl bromide (1 equiv.). The solution was stirred at room temperature for 12 h, then poured on water (100 mL) and extracted with Et$_2$O (3 × 50 mL). The combined organic phases were washed with water (25 mL) and dried with MgSO$_4$. Evaporation of the solvent in the vacuum gave the crude product which was purified as described below.

General Procedure B (GP B): A solution of the benzhydryl bromide in CH$_3$CN (10 mL) was prepared. After addition of amine (2.5 equiv), the solution was stirred at room temperature for 12 h, then poured on water (100 mL) and extracted with Et$_2$O (3 × 50 mL). The combined organic phases were washed with water (25 mL) and dried with MgSO$_4$.

3. Can One Predict Changes from S_N1 to S_N2 Mechanisms?

Evaporation of the solvent in the vacuum gave the crude product which was purified as described below.

(4,4'-Dichlorobenzhydryl)-propyl-amine (GP A) was obtained from 4,4'-dichlorobenzhydryl bromide (3.0 g, 9.5 mmol) and n-propylamine (11.2 g, 189 mmol). The crude product was distilled in the vacuum (127-137 °C/1 × 10^{-3} mbar): colorless oil (1.6 g, 57 %).

^{1}H NMR (300 MHz, CDCl$_3$): δ = 7.30 (d, J = 8.7 Hz, 4 H, ArH), 7.24 (d, J = 8.7 Hz, 4 H, ArH), 4.74 (s, 1 H, CHN), 2.49 (t, J = 6.9 Hz, 2 H), 1.51 (sext, J = 7.2 Hz, 2 H), 1.47 (br s, 1 H, NH), 0.90 (t, J = 7.2 Hz, 3 H);

^{13}C NMR (75.5 MHz, CDCl$_3$): δ = 142.5, 132.8, 128.7, 128.5, 66.3, 50.0, 23.3, 11.8;

MS (EI): m/z = 295, 293, 235; HR-MS (EI):

Calcd m/z for C$_{16}$H$_{17}$^{35}Cl$_2$N: 293.0740, Found: 293.0732.

***N*-(4,4'-Dichlorobenzhydryl)-piperidine** (GP A) was obtained from 4,4'-dichlorobenzhydryl bromide (3.5 g, 11 mmol) and piperidine (18.9 g, 221 mmol) after purification of the crude product by crystallization from methanol: colorless crystals (3.0 g, 85 %).

^{1}H NMR (300 MHz, CDCl$_3$): δ = 7.28 (d, J = 8.5 Hz, 4 H, ArH), 7.21 (d, J = 8.6 Hz, 4 H, ArH), 4.18 (s, 1 H), 2.26-2.25 (m, 4 H), 1.57–1.50 (m, 4 H), 1.44–1.38 (m, 2 H);

^{13}C NMR (75.5 MHz, CDCl$_3$): δ = 141.2, 132.5, 129.2, 128.6, 75.1, 52.9, 26.2, 24.5.

MS (EI): m/z = 321, 319, 235;

HR-MS (EI): Calcd. m/z for: C$_{18}$H$_{19}$^{35}Cl$_2$N: 319.0895, Found: 319.0885.

***N*-(4,4'-Dichlorbenzhydryl)-morpholine** (GP A) was obtained from 4,4'-dichlorobenzhydryl bromide (2.5 g, 7.9 mmol) and morpholine (13.8 g, 158 mmol) after purification of the crude product by crystallization from methanol: colorless crystals (2.17 g, 85 %).

^{1}H NMR (300 MHz, CDCl$_3$): δ 7.32 (d, J = 8.4 Hz, 4 H, ArH), 7.24 (d, J = 8.7 Hz, 4 H, ArH), 4.16 (s, 1 H), 3.71–3.68 (m, 4 H), 2.36–2.33 (m, 4 H);

^{13}C NMR (75.5 MHz, CDCl$_3$): δ = 140.4, 133.0, 129.1, 128.9, 77.1, 67.1, 52.5;

3. Can One Predict Changes from S_N1 to S_N2 Mechanisms?

MS (EI): m/z = 322, 323;

HR-MS (EI): Calcd m/z for for $C_{17}H_{17}{}^{35}Cl_2NO$: 321.0687, Found: 321.0680.

***N*-Benzhydryl-propyl-amine** (GP B) was obtained from benzhydryl bromide (3.0 g, 12 mmol) and *n*-propylamine (1.8 g, 30 mmol). The crude product was distilled in the vacuum (150-152 °C/2.1 × 10^{-2} mbar): colorless oil (1.85 g, 68 %).

^{1}H NMR (300 MHz, CDCl$_3$): δ = 7.39 (d, J = 7.4 Hz, 4 H), 7.27 (t, J = 7.2 Hz, 4 H), 7.17 (t, J = 7.0 Hz, 2 H), 4.80 (s, 1 H), 2.53 (t, J = 7.0 Hz, 2 H), 1.52 (sext, J = 7.3 Hz, 2 H), 1.46 (br s, 1 H, NH), 0.90 (s, J = 7.2 Hz, 3 H);

^{13}C NMR (75.5 MHZ, CDCl$_3$): δ = 144.4, 128.4, 127.2, 126.8, 67.5, 50.2, 23.3, 11.8;

MS (EI): m/z = 225, 196; HR-MS (EI): Calcd m/z for $C_{16}H_{19}N$: 225.1518, Found: 225.1504.

***N*-Benzhydryl-piperidine** (GP B) was obtained from benzhydryl bromide (2.5 g, 10 mmol) and piperidine (2.2 g, 26 mmol) after purification of the crude product by crystallization from methanol: colorless crystals (1.6 g, 64 %).

^{1}H NMR (300 MHz, CDCl$_3$): δ = 7.39 (d, J$_{HH}$ = 7.3 Hz, 4 H), 7.23 (t, J = 7.3 Hz, 4 H), 7.13 (t, J = 7.2 Hz, 2 H), 4.22 (s, 1 H), 2.35–2.28 (m, 4 H), 1.59-1.51 (m, 4 H), 1.45–1.37 (m, 2 H);

^{13}C NMR (75.5 MHz, CDCl$_3$): δ = 143.2, 128.3, 128.0, 126.6, 76.7, 53.1, 26.3, 24.7;

MS (EI): m/z = 252, 251, 174,167;

HR-MS (EI): Calcd m/z for $C_{18}H_{21}N$: 251.1674, Found: 251.1663.

***N*-Benzhydryl-morpholine** (GP B) was obtained from benzhydryl bromide (2.0 g, 8.1 mmol) and morpholine (1.8 g, 21 mmol) after purification of the crude product by crystallization from methanol: colorless crystals (1.6 g, 78 %).

^{1}H NMR (300 MHz, CDCl$_3$): δ =7.41 (d, J = 7.2 Hz, 4 H), 7.24 (t, J = 7.2 Hz, 4 H), 7.14 (t, J = 7.3 Hz, 2 H), 4.80 (s, 1 H), 3.70–3.67 (m, 4 H), 2.38–2.35 ppm (m, 4 H);

^{13}C NMR (75.5 MHz, CDCl$_3$): δ = 142.3, 128.5, 127.9, 126.9, 76.6, 67.1, 52.6 ppm;

MS (EI): m/z = 254, 253, 176, 167;

HR-MS (EI): Calcd m/z for $C_{16}H_{19}N$: 253.1467, Found: 253.1461.

3. Can One Predict Changes from S_N1 to S_N2 Mechanisms?

N-(4-Methylbenzhydryl)-propyl-amine (GP B) was obtained from 4-methylbenzhydryl bromide (1.5 g (5.7 mmol) and *n*-propylamine (0.85 g, 14 mmol) after purification of the crude product by chromatography (silica gel, 10/1 → 5/1 n-pentane/ethyl acetate, v/v): colorless oil (0.70 g, 51 %).
^{1}H NMR (300 MHz, CDCl$_3$): δ = 7.38 (d, J = 7.4 Hz, 2 H), 7.28-7.14 (m, 5 H), 7.08 (d, J = 7.8 Hz, 2 H), 4.77 (s, 1 H), 2.53 (t, J = 7.2 Hz, 2 H), 2.28 (s, 3 H, CH$_3$), 1.52 (sext, J = 7.2 Hz, 2 H), 1.46 (s, 1 H, NH), 0.90 (t, J = 7.2 Hz, 3 H);
^{13}C NMR (75.5 MHz, CDCl$_3$): δ = 144.7, 141.6, 136.5, 129.2, 128.5, 127.3, 127.2, 126.8, 67.3, 50.3, 23.4, 21.1, 11.9;
MS (EI): m/z = 240, 239, 181;
HR-MS (EI): Calcd. m/z for C$_{17}$H$_{21}$N: 239.1674, Found: 239.1670.

N-(4-Methylbenzhydryl)-piperidine (GP B) was obtained from 4-methylbenzhydryl bromide (1.5 g, 5.7 mmol) and piperidine (1.2 g, 14 mmol) after purification of the crude product by chromatography (silica gel, 15/1 n-pentane/ethyl acetate, v/v): colorless oil (0.79 g, 52 %).
^{1}H NMR (300 MHz, CDCl$_3$): δ = 7.46 (d, J = 7.5 Hz, 2 H), 7.36–7.18 (m, 5 H), 7.13 (d, J = 7.8 Hz, 2 H), 4.26 (s, 1 H), 2.38 (br t, 4 H), 2.34 (br s, 3 H), 1.66–1.59 (m, 4 H), 1.52-1.45 ppm (m, 2 H);
^{13}C NMR (75.5 MHz, CDCl$_3$): δ 143.5, 140.2, 136.1, 129.0, 128.3, 127.9, 126.5, 76.4, 53.1, 26.3, 24.7, 21.0 ppm;
MS (EI): m/z = 266, 265, 181;
HR-MS (EI): Calcd. m/z for C$_{19}$H$_{23}$N: 265.1831, Found: 265.1828.

N-(4-Methylbenzhydryl)-morpholine (GP B) was obtained from 4-methylbenzhydryl bromide (1.5 g, 5.7 mmol) and morpholine (1.3 g, 15 mmol) after purification of the crude product by chromatography (silica gel, 20/1 *n*-pentane/ethyl acetate, v/v): colorless oil (1.1 g, 72 %).
^{1}H NMR (300 MHz, CDCl$_3$): δ = 7.41 (d, J = 7.5 Hz, 2 H), 7.40–7.12 (m, 5 H), 7.07 (d, J = 7.8 Hz, 2 H), 4.15 (s, 1 H), 3.69 (br t, J = 4.6 Hz, 4 H), 2.37 (br t, J = 4.2 Hz, 4 H), 2.26 ppm (s, 3 H);

3. Can One Predict Changes from S_N1 to S_N2 Mechanisms?

^{13}C NMR (75.5 MHz, CDCl$_3$): δ = 142.5, 139.3, 136.5, 129.2, 128.5, 127.8, 127.8, 126.9, 76.4, 67.1, 52.6, 21.0 ppm.

Benzhydryl Methyl Ether was synthesized in analogy to a procedure by Olah and Welch.[43] Benzhydrol (10 g) was dissolved in methanol (120 mL). After addition of a catalytic amount of trifluoromethansulfonic acid (approximately 0.5 mL) the mixture was heated to reflux for 2.5 h. Then the solvent was removed under reduced pressure. The resulting oil was poured on water (100 mL) and a spatula tip of K$_2$CO$_3$ was added. The mixture was extracted with 1/1 n-pentane/diethylether (3 × 50 mL). The combined organic phases were dried over MgSO$_4$. Evaporation of the solvent gave the crude product as a colorless oil. For purification the product was distilled from sodium in the vacuum (83-84 °C/1 × 10^{-3} mbar) to give a colorless oil (7.8 g, 72 %). ^{1}H NMR (300 MHz, CDCl$_3$): δ = 7.34-7.16 (m, 10 H, ArH), 5.20 (s, 1 H, CHOMe), 3.33 (s, 3 H, Me);
^{13}C NMR (75.5 MHz, CDCl$_3$): δ = 142.0, 128.3, 126.8, 127.3, 85.3, 56.8;
MS (EI): m/z = 199, 198 [M$^+$], 167;
HR-MS (EI): Calcd. m/z for C$_{14}$H$_{14}$O: 198.1045, Found: 198.1056.

3. Can One Predict Changes from S_N1 to S_N2 Mechanisms?

Method

The product studies were carried out for several representative systems to examine the ratios of the products formed during the reactions of benzhydryl bromides (1-X-Y) with 0.2 M solutions of amines in DMSO.

The concentrations of the components in the reaction mixtures were determined by GC with a flame ionization detector (FID) using n-hexadecane (C16) as an internal standard. The ratio of the peak area of a compound (A_{Cpd}) to the peak area of a given standard (A_{Std}), is not equal to the ratio of the molar amount of the compound [Cpd] to the molar amount of the standard [Std]. Therefore, the specific sensitivity of the FID for different molecules was accounted for by defining relative molecular response (RMR) factors (eq. 3.7) for each of the possible products which were synthesized independently.

$$\frac{[Cpd]}{[Std]} = RMR \frac{A_{Cpd}}{A_{Std}} \qquad (3.7)$$

The retention times (t_R in min) and the RMR values for each product with respect to hexadecane (C16 = Std) were determined from the peak areas in chromatograms with known [Cpd] and [Std] (eq. 3.8).

$$RMR = \frac{[Cpd]A_{Std}}{[Std]A_{Cpd}} \qquad (3.8)$$

Thus, the product concentrations tabulated were calculated according to equation 3.6 in which A_{Cpd} and A_{Std} are the peak areas of the compound and the standard C16, respectively, and [Std] is the known concentration of the internal standard C16.

$$[Cpd] = RMR[Std]\frac{A_{Cpd}}{A_{Std}} \qquad (3.6)$$

3. Can One Predict Changes from S_N1 to S_N2 Mechanisms?

Standardization

As defined in equation 3.8 the relative molecular response (RMR) factors of benzophenone, benzhydrol, and several representative benzhydryl amines was determined by GC with respect to n-hexadecane (C_{16}) used as an internal standard.

Table 3.13. Determination of the relative molecular response factors (RMR) for **10-OH, 10=O** and **10=NR$_2$**.

	c/mol L^{-1}	average A	RMR	t_R/min	average RMR
C16	1.46 × 10^{-3}	18252107		5.20	
4-methylbenzhydrol	6.21 × 10^{-3}	59502760	1.31	9.84	1.29 [a]
C16	1.46 × 10^{-3}	17548737			
4-methylbenzhydrol	1.24 × 10^{-2}	116746533	1.28		
C16	1.46 × 10^{-3}	17516347			
4-methylbenzhydrol	1.86 × 10^{-2}	173877367	1.29		
C16	1.46 × 10^{-3}	17991280			
4-methylbenzophenone	5.72 × 10^{-3}	55033793	1.28	9.96	1.29 [a]
C16	1.46 × 10^{-3}	18125710			
4-methylbenzophenone	1.14 × 10^{-2}	110819800	1.29		
C16	1.46 × 10^{-3}	17433973			
4-methylbenzophenone	1.72 × 10^{-2}	159565800	1.29		
C16	1.46 × 10^{-3}	18392697		5.22	
N-(4-methylbenzhydryl)morpholine	1.85 × 10^{-3}	22421527	1.04	13.1	1.07

[a] In the case of 4-methylbenzophenone (**10=O**) and 4-methylbenzhydrol (**10-OH**) the retention time and RMR value were almost identical.

3. Can One Predict Changes from S_N1 to S_N2 Mechanisms?

Table 3.13. (continued)

	c/mol L^{-1}	average A	RMR	t_R/min	average RMR
C16	1.46×10^{-3}	17689997			
N-(4-methylbenzhydryl)morpholine	7.41×10^{-3}	82840557	1.09		
C16	1.46×10^{-3}	17577083			
N-(4-methylbenzhydryl)morpholine	1.48×10^{-2}	159630267	1.12		
C16	1.46×10^{-3}	17779833			
N-(4-methylbenzhydryl)morpholine	3.70×10^{-3}	44007437	1.03		
C16	1.46×10^{-3}	18424133		5.22	
N-(4-methylbenzhydryl)-propyl-amine	2.69×10^{-3}	33041727	1.03	10.14	1.07
C16	1.46×10^{-3}	17794760			
N-(4-methylbenzhydryl)-propyl-amine	5.38×10^{-3}	61645867	1.07		
C16	1.46×10^{-3}	18698703			
N-(4-methylbenzhydryl)-propyl-amine	1.08×10^{-2}	127503300	1.08		
C16	1.46×10^{-3}	18647680			
N-(4-methylbenzhydryl)-propyl-amine	1.61×10^{-2}	185498700	1.11		

3. Can One Predict Changes from S_N1 to S_N2 Mechanisms?

Table 3.13. (continued)

	c/mol L^{-1}	average A	RMR	t_R/min	average RMR
C16	1.46×10^{-3}	17841197		5.22	
N-(4-methylbenzhydryl)piperidine	1.78×10^{-3}	22589587	0.96	12.24	1.06
C16	1.46×10^{-3}	18263093			
N-(4-methylbenzhydryl)piperidine	3.70×10^{-3}	40523610	1.15		
C16	1.46×10^{-3}	18352357			
N-(4-methylbenzhydryl)piperidine	7.41×10^{-3}	88217353	1.06		
C16	1.46×10^{-3}	17164033			
N-(4-methylbenzhydryl)piperidine	1.48×10^{-2}	165438033	1.06		

Table 3.14. Determination of the relative molecular response factors (RMR) for **9-OH**, **9=O** and **9=NR$_2$**.

	c/mol L^{-1}	average A	RMR	t_R/min	average RMR
C16	1.08×10^{-3}	12616997		5.18	
benzhydrol	5.13×10^{-4}	5118717	1.17	8.34	1.17
benzophenone	2.36×10^{-3}	21475193	1.28	7.98	1.34
N-benzhydrylpiperidine	1.35×10^{-3}	18003217	0.88	11.12	0.90
C16	1.08×10^{-3}	12787997			
benzhydrol	1.03×10^{-3}	10627827	1.14		
benzophenone	5.90×10^{-4}	5126065	1.36		
N-benzhydrylpiperidine	4.51×10^{-4}	5675417	0.94		

3. Can One Predict Changes from S_N1 to S_N2 Mechanisms?

Table 3.14. (continued)

	c/mol L^{-1}	average A	RMR	t_R/min	average RMR
C16	1.08×10^{-3}	13296300			
benzhydrol	1.54×10^{-3}	16139817	1.17		
benzophenone	1.18×10^{-3}	10621647	1.37		
N-benzhydrylpiperidine	9.02×10^{-4}	12755333	0.87		
C16	1.08×10^{-3}	13162683			
benzhydrol	2.56×10^{-4}	2615610	1.19		
benzophenone	1.89×10^{-3}	17007627	1.35		
C16	1.09×10^{-3}	12668245		5.18	
N-benzhydryl-propyl-amine	7.55×10^{-3}	76862430	1.14	9.12	1.11
C16	1.09×10^{-3}	13187900			
N-benzhydryl-propyl-amine	1.13×10^{-2}	117953400	1.16		
C16	1.09×10^{-3}	13471903			
N-benzhydryl-propyl-amine	5.66×10^{-3}	63520567	1.10		
C16	1.09×10^{-3}	13104907			
N-benzhydryl-propyl-amine	1.89×10^{-3}	22196133	1.02		
C16	1.09×10^{-3}	13028710		5.18	
N-benzhydrylmorpholine	1.47×10^{-3}	16938615	1.04	12.00	1.04
C16	1.09×10^{-3}	13394560			
N-benzhydrylmorpholine	2.21×10^{-3}	25147613	1.08		

3. Can One Predict Changes from S_N1 to S_N2 Mechanisms?

Table 3.14. (continued)

	c/mol L−1	average A	RMR	t_R/min	average RMR
C16	1.09 × 10⁻³	12996903			
N-benzhydrylmorpholine	1.10 × 10⁻³	13195633	1.00		
C16	1.09 × 10⁻¹	13028693			
N-benzhydrylmorpholine	5.15 × 10⁻³	59620750	1.03		
C16	1.16 × 10⁻³	14734243		5.22	
benzhydryl methyl ether	4.64 × 10⁻³	59440763	0.99	6.94	1.02
C16	1.16 × 10⁻³	15104223			
benzhydryl methyl ether	9.27 × 10⁻³	119391333	1.01		
C16	1.16 × 10⁻³	14696190			
benzhydryl methyl ether	1.39 × 10⁻²	172305233	1.02		
C16	1.16 × 10⁻³	15661193			
benzhydryl methyl ether	1.85 × 10⁻²	239973200	1.04		

Table 3.15. Determination of the relative molecular response factors (RMR) for **8-OH, 8=O** and **8=NR₂**.

	c/mol L⁻¹	average A	RMR	t_R/min	average RMR
C16	1.13 × 10⁻	13573363		5.18	
4,4'-dichlorobenzhydrol	1.23 × 10⁻³	9333182	1.59	14.38	1.52
4,4'-dichlorobenzophenone	2.49 × 10⁻³	19730167	1.52	12.68	1.45

3. Can One Predict Changes from S_N1 to S_N2 Mechanisms?

Table 3.15. (continued)

	c/mol L^{-1}	average A	RMR	t_R/min	average RMR
C16	1.13 × 10^{-3}	13929460			
4,4'-dichlorobenzhydrol	2.47 × 10^{-3}	20626030	1.48		
4,4'-dichlorobenzophenone	4.99 × 10^{-3}	42420953	1.45		
C16	1.13 × 10^{-3}	13556390			
4,4'-dichlorobenzhydrol	4.94 × 10^{-3}	40377470	1.47		
4,4'-dichlorobenzophenone	1.25 × 10^{-3}	10769563	1.39		
C16	1.13 × 10^{-3}	12544883			
4,4'-dichlorobenzhydrol	9.88 × 10^{-3}	74436677	1.48		
4,4'-dichlorobenzophenone	4.99 × 10^{-3}	39232363	1.42		
C16	1.13 × 10^{-3}	13118177			
4,4'-dichlorobenzhydrol	7.41 × 10^{-3}	55738440	1.55		
4,4'-dichlorobenzophenone	9.98 × 10^{-3}	78744740	1.48		
C16	1.13 × 10^{-3}	12947747			
4,4'-dichlorobenzhydrol	4.94 × 10^{-3}	37237693	1.52		
4,4'-dichlorobenzophenone	7.48 × 10^{-3}	59802570	1.44		
C16	1.08 × 10^{-3}	12483880		5.18	
N-(4,4'-dichlorobenzhydryl)morpholine	1.34 × 10^{-3}	13539817	0.94	17.14	0.96
C16	1.08 × 10^{-3}	13943160			
N-(4,4'-dichlorobenzhydryl)morpholine	1.78 × 10^{-3}	19422840	0.99		

3. Can One Predict Changes from S_N1 to S_N2 Mechanisms?

Table 3.15. (continued)

	c/mol L^{-1}	average A	RMR	t_R/min	average RMR
C16	1.08×10^{-3}	12387687			
N-(4,4'-dichlorobenzhydryl)morpholine	1.51×10^{-3}	15393007	0.90		
C16	1.08×10^{-3}	13861360			
N-(4,4'-dichlorobenzhydryl)morpholine	8.91×10^{-4}	9667222	1.00		
C16	9.30×10^{-4}	13933403		5.20	
N-(4,4'-dichlorobenzhydryl)-propylamine	7.44×10^{-3}	79286523	1.41	14.48	1.41
C16	9.30×10^{-4}	13766470			
N-(4,4'-dichlorobenzhydryl)-propylamine	1.12×10^{-2}	113108667	1.46		
C16	9.30×10^{-4}	13741350			
N-(4,4'-dichlorobenzhydryl)-propylamine	3.72×10^{-3}	40173987	1.37		
C16	1.16×10^{-3}	14842043		5.18	
N-(4,4'-dichlorobenzhydryl)piperidine	6.09×10^{-3}	70046967	1.11	16.52	1.14
C16	1.16×10^{-3}	17969947			
N-(4,4'-dichlorobenzhydryl)piperidine	2.03×10^{-3}	28178667	1.12		
C16	1.16×10^{-3}	15486927			
N-(4,4'-dichlorobenzhydryl)piperidine	4.06×10^{-3}	48926543	1.11		
C16	1.16×10^{-3}	14583050			
N-(4,4'-dichlorobenzhydryl)piperidine	3.05×10^{-3}	31454863	1.22		

3. Can One Predict Changes from S_N1 to S_N2 Mechanisms?

Analysis of Product Mixtures

A 0.2 M solution of the particular amine in DMSO (25 mL) was prepared in a volumetric flask and kept at 20 °C. Then this solution was added to the particular benzhydryl bromide (0.5 mmol) and stirred for 24 h. After addition of water (50 mL) the solution was extracted with Et$_2$O (3 × 50 mL). The combined organic phases were dried (MgSO$_4$). After evaporation of the solvent the remaining oil was dissolved in acetone (ca. 20 mL) and transferred into a 25 mL-volumetric flask. Then a defined amount of an acetone stock solution containing the hexadecane standard (C16) was added, to give a C16 concentration of about 1 mM. The volumetric flask was then filled up to 25 mL with acetone, and aliquots were subjected to GC analysis (t_R = retention time; peak areas are the average of 3 separate GC runs for each sample). Concentrations of N°-OH, N°=O and N°=NR$_2$ were calculated by using equation 3.6 and refer to the known concentration of the standard (C16).

Table 3.16. Product analysis for the reaction of **10-Br** in the presence of 0.2 M of amine.

RR'NH	compound	t_R/min	peak area A	RMR	c/mol L^{-1}	[**10-NR$_2$**] /([**10=O**]+[**10-OH**])
piperidine	C16	5.20	14371033		1.16 × 10^{-3}	
	10-OH + **10=O**	9.84/9.96	67721127	1.29	7.08 × 10^{-3}	
	10-NR$_2$	12.56	97602543	1.07	8.46 × 10^{-3}	
						1.19
morpholine	C16	5.20	14378403		1.11 × 10^{-3}	
	10-OH + **10=O**	9.84/9.96	99732157	1.29	9.96 × 10^{-3}	
	10-NR$_2$	13.32	65561753	1.07	5.43 × 10^{-3}	
						0.54
n-PrNH$_2$	C16	5.20	13121633		1.05 × 10^{-3}	
	10-OH + **10=O**	9.84/9.96	156201167	1.29	1.61 × 10^{-2}	
	10-NR$_2$	10.32	34093090	1.07	2.92 × 10^{-3}	
						0.18

3. Can One Predict Changes from S_N1 to S_N2 Mechanisms?

Table 3.17. Product analysis for the reaction of **9-Br** in the presence of 0.2 M of amine.

RR'NH	compound	t_R/min	peak area A	RMR	c/mol L^{-1}	[9-NR$_2$] /([9=O]+[9-OH])
piperidine	C16	5.20	14843277		**1.16 × 10^{-3}**	
	9-OH	8.34	1263792	1.17	1.16 × 10^{-4}	
	9=O	7.98	14133813	1.34	1.49 × 10^{-3}	
	9-NR$_2$	11.6	198575867	0.90	1.40 × 10^{-2}	
						8.71
morpholine	C16	5.20	13262553		**1.04 × 10^{-3}**	
	9-OH	8.34	5840434	1.17	5.34 × 10^{-4}	
	9=O	7.98	30709390	1.34	3.22 × 10^{-3}	
	9-NR$_2$	12.48	172224433	1.04	1.40 × 10^{-2}	
						3.72
n-PrNH$_2$	C16	5.20	13528813		**1.04 × 10^{-3}**	
	9-OH	8.34	36709970	1.17	3.32 × 10^{-3}	
	9=O	7.98	51315043	1.34	5.33 × 10^{-3}	
	9-NR$_2$	9.12	113360600	1.11	9.67 × 10^{-3}	
						1.12

3. Can One Predict Changes from S_N1 to S_N2 Mechanisms?

Table 3.18. Product analysis for the reaction of **9-Br** in the presence of 0.2 M 2,6-Lutidine.

9-Br $\xrightarrow{\text{0.2 M 2,6-lutidine, DMSO, 20°C}}$ 9=O + 9-OH

rxn time/min	compound	t_R/min	peak area A	RMR	c/mol L^{-1}	[9=O]/[9-OH]
60	C16	5.20	13253270		1.13×10^{-3}	
	9=O	7.98	37614477	1.34	4.24×10^{-3}	
	9-OH	8.34	112488067	1.17	1.11×10^{-2}	0.38
120	C16	5.20	12744147		1.13×10^{-3}	
	9=O	7.98	74312043	1.34	8.71×10^{-3}	
	9-OH	8.34	77821767	1.17	7.95×10^{-3}	1.10
180	C16	5.20	13204473		1.13×10^{-3}	
	9=O	7.98	92219953	1.34	1.04×10^{-2}	
	9-OH	8.34	63360827	1.17	6.25×10^{-3}	1.67
300	C16	5.20	13204473		1.13×10^{-3}	
	9=O	7.98	92219953	1.34	1.26×10^{-2}	
	9-OH	8.34	63360827	1.17	3.75×10^{-3}	3.36
360	C16	5.20	13204473		1.13×10^{-3}	
	9=O	7.98	92219953	1.34	1.09×10^{-2}	
	9-OH	8.34	63360827	1.17	2.76×10^{-3}	3.95

In order to prove the intermediacy of the oxysulfonium ion N°-OS$^+$Me$_2$, the reaction of **9-Br** in DMSO containing 0.2 M 2,6-Lutidin was quenched with methanol after 6.5 h. Thus, 100 mL methanol were added and after stirring for 24 h the solution was worked-up as described above.

3. Can One Predict Changes from S_N1 to S_N2 Mechanisms?

Table 3.18. (continued)

rxn time/min	compound	t_R/min	peak area A	RMR	c/mol L^{-1}	[9=O]/ ([9-OH]+[9-OMe])
390	C16	5.20	13253270		1.13×10^{-3}	
	9=O	7.98	37614477	1.34	1.85×10^{-2}	
	9-OH	8.34	112488067	1.17	1.64×10^{-3}	
	9-OMe	6.42	29892213	1.02	4.27×10^{-3}	3.13

Table 3.19. Product analysis for the reaction of **8-Br** in the presence of 0.2 M of amine.

RR'NH	T/°C	compound	t_R/min	peak area A	RMR	c/mol L^{-1}	[8-NR$_2$] /([8=O]+[8-OH])
piperidine	20	C16	5.20	14845483		1.16×10^{-3}	
		8-OH	14.38	1001725	1.52	1.19×10^{-4}	
		8=O	12.68	2615282	1.45	2.98×10^{-4}	
		8-NR$_2$	16.52	172146633	1.14	1.54×10^{-2}	
							36.7
morpholine	20	C16	5.20	12987620		1.04×10^{-3}	
		8-OH	14.38	3263352	1.52	3.95×10^{-4}	
		8=O	12.68	4546383	1.45	5.26×10^{-4}	
		8-NR$_2$	17.62	132638233	0.96	1.01×10^{-2}	
							11.0
	35	C16	5.20	12923807		1.13×10^{-3}	
		8-OH	14.38	1700768	1.52	2.26×10^{-4}	
		8=O	12.68	9956992	1.45	1.27×10^{-3}	
		8-NR$_2$	17.62	126759467	0.96	1.07×10^{-2}	
							7.13

3. Can One Predict Changes from S_N1 to S_N2 Mechanisms?

Table 3.19. (continued)

RR'NH	T/°C	compound	t_R/min	peak area A	RMR	c/mol L^{-1}	[8-NR$_2$] /([8=O]+[8-OH])
	50	C16	5.20	13748827		**1.13 × 10^{-3}**	
		8-OH	14.38	1862981	1.52	2.33 × 10^{-4}	
		8=O	12.68	15290797	1.45	1.83 × 10^{-3}	
		8-NR$_2$	17.62	107793400	0.96	8.52 × 10^{-3}	
							4.13

3.6. References

(1) a) Streitwieser, A., Jr. *Solvolytic Displacement Reactions*; McGraw-Hill: New York, **1962**. b) Ingold, C. K. *Structure and Mechanism in Organic Chemistry*, 2nd ed.; Cornell University Press: Ithaca, NY, **1969**. c) Vogel, P. *Carbocation Chemistry*; Elsevier: Amsterdam, 1985. d) *Advances in Carbocation Chemistry*; Coxon, J. M., Ed.; JAI Press: Greenwich, CT, **1995**; Vol 2. e) Raber, D. J.; Harris, J. M.; Schleyer, P. v. R. In *Ions and Ion Pairs in Organic Reactions*; Szwarc, M., Ed.; Wiley: New York, **1974**; Vol. 2, pp 247-374. f) Smith, M. B.; March, J. *March's Advanced Organic Chemistry*, 6th ed.; Wiley: New York, **2007**; Chapter 10. g) Vlasov, V. M. *Russ. Chem. Rev.* **2006**, *75*, 765-796.

(2) a) Guthrie, R. D. *Pure Appl. Chem.* **1988**, *61*, 23-56. b) Guthrie, R. D.; Jencks, W. P. *Acc. Chem. Res.* **1989**, *22*, 343-349.

(3) Winstein, S.; Grunwald, E.; Jones, H. W. *J. Am. Chem. Soc.* **1951**, *73*, 2700-2707.

(4) a) Winstein, S.; Clippinger, E.; Fainberg, A. H.; Heck, R.; Robinson, G. C. *J. Am. Chem. Soc.* **1956**, *78*, 328-335. For a summary of Winstein's contribution, see: b) Bartlett, P. D. *J. Am. Chem. Soc.* **1972**, *94*, 2161-2170. c) Winstein, S.; Appel, B.; Baker, R.; Diaz, A. In *Symposium on Organic Reaction Mechanisms*, Chemical Society (London), Special Publication No. 19, **1965**; p 109.

3. Can One Predict Changes from S_N1 to S_N2 Mechanisms?

(5) a) Sneen, R. A. *Acc. Chem. Res.* **1973**, *6*, 46-53. b) Sneen, R. A.; Larsen, J. W. *J. Am. Chem. Soc.* **1969**, *91*, 6031-6035. c) Sneen, R. A.; Larsen, J. W. *J. Am. Chem. Soc.* **1969**, *91*, 362-366.

(6) a) Bentley, T. W.; Schleyer, P. v. R. *J. Am. Chem. Soc.* **1976**, *98*, 7658-7666. b) Schadt, F. L.; Bentley, T. W.; Schleyer, P. v. R. *J. Am. Chem. Soc.* **1976**, *98*, 7667-7675. c) Bentley, T. W.; Bowen, C. T.; Morten, D. H.; Schleyer, P. v. R. *J. Am. Chem. Soc.* **1981**, *103*, 5466-5475.

(7) Bentley, T. W.; Schleyer, P. v. R. *Adv. Phys. Org. Chem.* **1977**, *14*, 1-67.

(8) a) Gold, V. *J. Chem. Soc.* 1956, 4633-4637. b) Gregory, B. J.; Kohnstam, G.; Paddon-Row, M.; Queen, A. *J. Chem. Soc. D* **1970**, 1032-1033. c) Gregory, B. J.; Kohnstam, G.; Queen, A.; Reid, D. J. *J. Chem. Soc., Chem. Commun.* **1971**, 797-799. d) Buckley, N.; Oppenheimer, N. J. *J. Org. Chem.* **1997**, 62, 540-551.

(9) a) Ceccon, A.; Papa, I.; Fava, A. *J. Am. Chem. Soc.* **1966**, *88*, 4643-4648. b) Fava, A.; Iliceto, A.; Ceccon, A. *Tetrahedron Lett.* **1963**, 685-692.

(10) a) Diaz, A. F.; Winstein, S. *J. Am. Chem. Soc.* **1964**, *86*, 5010-5011. b) Casapieri, P.; Swart, E. R. *J. Chem. Soc.* **1963**, 1254-1262. c) Casapieri, P.; Swart, E. R. *J. Chem. Soc.* **1961**, 4342-4347.

(11) a) Pocker, Y. *J. Chem. Soc.* **1959**, 3939-3943. b) Pocker, Y. *J. Chem. Soc.* **1959**, 3944-3949.

(12) a) Lim, C.; Kim, S.-H.; Yoh, S.-D.; Fujio, M.; Tsuno, Y. *Tetrahedron Lett.* **1997**, *38*, 3243-3246. b) Tsuno, Y.; Fujio, M. *Adv. Phys. Org. Chem.* **1999**, *32*, 267-385.

(13) a) Kim, S. H.; Yoh, S.-D.; Lim, C.; Mishima, M.; Fujio, M.; Tsuno, Y. *J. Phys. Org. Chem.* **1998**, *11*, 254-260. b) Kim, S. H.; Yoh, S.-D.; Fujio, M.; Imahori, H.; Mishima, M.; Tsuno, Y. *Bull. Korean Chem. Soc.* **1995**, *16*, 760-764. c) Yoh, S.-D.; Tsuno, Y.; Fujio, M.; Sawada, M.; Yukawa, Y. *J. Chem. Soc., Perkin Trans. 2* **1989**, 7-13. d) Yoh, S. D. *J. Korean Chem. Soc.* **1975**, *19*, 240–245.

(14) a) Yoh, S.-D.; Cheong, D.-Y.; Lee, C.-H.; Kim, S.-H.; Park, J.-H.; Fujio, M.; Tsuno, Y. *J. Phys. Org. Chem.* **2001**, *14*, 123-130. b) Yoh, S. D.; Lee, M.-K.; Son, K.-J.; Cheong, D.-Y.; Han, I.-S.; Shim, K.-T. *Bull. Korean Chem. Soc.* **1999**, *20*, 466-468. c) Amyes, T. L.; Richard, J. P. *J. Am. Chem. Soc.* **1990**, *112*, 9507-9512.

3. Can One Predict Changes from S_N1 to S_N2 Mechanisms?

(15) a) Katritzky, A. R.; Brycki, B. E. *Chem. Soc. Rev.* **1990**, *19*, 83-105. b) Katritzky, A. R.; Brycki, B. E. *J. Phys. Org. Chem.* **1988**, *1*, 1-20. c) Katritzky, A. R.; Musumarra, G. *Chem. Soc. ReV.* **1984**, *13*, 47-68. d) Katritzky, A. R.; Sakizadeh, K.; Gabrielsen, B.; le Noble, W. J. *J. Am. Chem. Soc.* **1984**, *106*, 1879-1880. e) Katritzky, A. R.; Musumarra, G.; Sakizadeh, K. *J. Org. Chem.* **1981**, *46*, 3831-3835.

(16) a) Jencks, W. P. *Acc. Chem. Res.* **1980**, *13*, 161-169. b) Jencks, W. P. *Chem. Soc. Rev.* **1981**, *10*, 345-375. c) Richard, J. P.; Jencks, W. P. *J. Am. Chem. Soc.* **1984**, *106*, 1373-1383. d) Richard, J. P.; Jencks, W. P. *J. Am. Chem. Soc.* **1984**, *106*, 1383-1396.

(17) a) Richard, J. P. In *Advances in Carbocation Chemistry*; Creary, X., Ed.; JAI Press: Greenwich, CT, **1989**; Vol. 1, pp 121-169. b) Amyes, T. L.; Toteva, M. M.; Richard, J. P. In *Reactive Intermediate Chemistry*; Moss, R. A., Platz, M. S., Jones, M., Jr., Eds.; Wiley-Interscience: Hoboken, NJ, **2004**; pp 41-69. c) Richard, J. P.; Amyes, T. L.; Toteva, M. M.; Tsuji, Y. *AdV. Phys. Org. Chem.* **2004**, *39*, 1-26.

(18) a) Database of reactivity parameters E, N, and s_N: http://www.cup.uni-muenchen.de/oc/mayr. b) Mayr, H.; Bug, T.; Gotta, M. F.; Hering, N.; Irrgang, B.; Janker, B.; Kempf, B.; Loos, R.; Ofial, A. R.; Remennikov, G.; Schimmel, H. *J. Am. Chem. Soc.* **2001**, *123*, 9500-9512.

(19) a) Mayr, H.; Ofial, A. R. *Pure Appl. Chem.* **2005**, *77*, 1807-1821. b) Mayr, H.; Ofial, A. R. *J. Phys. Org. Chem.* **2008**, *21*, 584–595. c) Mayr, H.; Patz, M. *Angew. Chem.* **1994**, *106*, 990-1010; *Angew. Chem., Int. Ed. Engl.* **1994**, *33*, 938-957.

(20) Mayr, H.; Kempf, B.; Ofial, A. R. *Acc. Chem. Res.* **2003**, *36*, 66-77.

(21) a) Denegri, B.; Ofial, A. R.; Juric, S.; Streiter, A.; Kronja, O.; Mayr, H. *Chem. Eur. J.* **2006**, *12*, 1657-1666. b) Denegri, B.; Streiter, A.; Juric, S.; Ofial, A. R.; Kronja, O.; Mayr, H. *Chem. Eur. J.* **2006**, *12*, 1648-1656.

(22) Creary, X.; Burtch, E. A. *J. Org. Chem.* **2004**, *69*, 1227-1234.

(23) Hansch, C.; Leo, A.; Taft, R. W. *Chem. Rev.* **1991**, *91*, 165-195.

(24) Uddin, K.; Fujio, M.; Kim, H.-J.; Rappoport, Z.; Tsuno, Y. *Bull. Chem. Soc. Jpn.* **2002**, *75*, 1371-1379.

3. Can One Predict Changes from S_N1 to S_N2 Mechanisms?

(25) Hammett plots of the second-order rate constants of the reactions of substituted benzyl bromides with pyridine (20 °C, in acetone) are shown in the Supporting Information: a) Baker, J. W. *J. Chem. Soc.* **1936**, 1448–1451. b) Baker, J. W. *Trans. Faraday Soc.* **1941**, *37*, 632–644.

(26) Young, P. R.; Jencks, W. P. *J. Am. Chem. Soc.* **1979**, *101*, 3288-3294.

(27) a) Bordwell, F. G.; Hughes, D. L. *J. Org. Chem.* **1980**, *45*, 3320-3325. b) Bordwell, F. G.; Hughes, D. L. *J. Am. Chem. Soc.* **1986**, *108*, 7300-7309.

(28) a) Kornblum, N.; Jones, W. J.; Anderson, G. J. *J. Am. Chem. Soc.* **1959**, *81*, 4113-4114. b) Kornblum, N.; Powers, J. W.; Anderson, G. J.; Jones, W. J.; Larson, H. O.; Levand, O.; Weaver, W. M. *J. Am. Chem. Soc.* **1957**, *79*, 6562-6562. c) Dave, P.; Byun, H.-S.; Engel, R. *Synth. Commun.* **1986**, *16*, 1343-1346.

(29) a) Torssell, K. *Tetrahedron Lett.* **1966**, *7*, 4445-4451. b) Torssell, K. *Acta Chem. Scand.* **1967**, *21*, 1-14.

(30) Creary, X.; Burtch, E. A.; Jiang, Z. *J. Org. Chem.* **2003**, *68*, 1117-1127. b) Duty, R. C.; Gurne, R. L. *J. Org. Chem.* **1970**, *35*, 1800-1802.

(31) a) Arnett, E. M.; Reich, R. *J. Am. Chem. Soc.* **1980**, *102*, 5892-5902. b) Duty, R. C.; Gurne, R. L. *J. Org. Chem.* **1970**, *35*, 1800-1802.

(32) a) Cowie, G. R.; Fitches, H. J. M.; Kohnstam, G. *J. Chem. Soc.* **1963**, 1585–1593. b) Fox, J. R.; Kohnstam, G. *J. Chem. Soc.* **1963**, 1593-1598.

(33) Brotzel, F.; Chu, Y. C.; Mayr, H. *J. Org. Chem.* **2007**, *72*, 3679-3688.

(34) Kanzian, T.; Nigst, T. A.; Maier, A.; Pichl, S.; Mayr, H. *Eur. J. Org. Chem.* **2009**, 6379-6385.

(35) Phan, T. B.; Breugst, M.; Mayr, H. *Angew. Chem.* **2006**, *118*, 3954-3959; *Angew. Chem., Int. Ed.* **2006**, *45*, 3869-3874.

(36) Minegishi, S.; Mayr, H. *J. Am. Chem. Soc.* **2003**, *125*, 286-295.

(37) a) Ho, T.-L. *Hard and Soft Acids and Bases Principle in Organic Chemistry*; Academic Press: New York, **1977**. b) Smith, S. G.; Winstein, S. *Tetrahedron* **1958**, *3*, 317-318. c) Rasul, G.; Prakash, G. K. S.; Olah, G. A. *J. Org. Chem.* **2000**, *65*, 8786-8789. d) da Silva, R. R.; Santos, J. M.; Ramalho, T. C.; Figueroa-Villar, J. D. *J. Braz. Chem. Soc.* **2006**, *17*, 223-226.

3. Can One Predict Changes from S_N1 to S_N2 Mechanisms?

(38) McClelland, R.; Kanagasabapathy, V. M.; Banait, N. S.; Steenken, S. *J. Am. Chem. Soc.* **1989**, *111*, 3966-3972.

(39) Minegishi, S.; Kobayashi, S.; Mayr, H. *J. Am. Chem. Soc.* **2004**, *126*, 5174-5181.

(40) Measurements of the nucleophilic reactivity of DMSO in neat DMSO are not possible with our equipment because the laser radiation at 266 nm, which is needed for the photoionization of the benzhydryl chlorides, is absorbed by DMSO.

(41) Baidya, M.; Kobayashi, S.; Brotzel, F.; Schmidhammer, U.; Riedle, E.; Mayr, H. *Angew. Chem.* **2007**, *119*, 6288-6292; *Angew. Chem. Int. Ed.* **2007**, *46*, 6176-6179.

(42) Phan, T. B.; Nolte, C.; Kobayashi, S.; Ofial, A. R.; Mayr, H. *J. Am. Chem. Soc.* **2009**, 131, 11392-11401.

(43) Olah, G. A.; Welch, J. *J. Am. Chem. Soc.* **1978**, *100*, 5396-5402.

4. Leaving Group Dependence of the S_N1/S_N2 Ratio

4.1. Introduction

In the preceding chapter the change from S_N1 to S_N2 mechanism has only been investigated for benzhydryl bromides. In this chapter, the methodology of the previous chapter will be used to investigate the change from S_N1 to S_N2 mechanism for the *meta*-fluorinated benzhydryl bromides (**1,3,4,7**)-Br and tosylates (**1,3,4,7**)-OTs introduced in Chapter 1.

Scheme 4.1. Benzhydryl derivatives employed in this study; E_f parameters are given in parentheses.

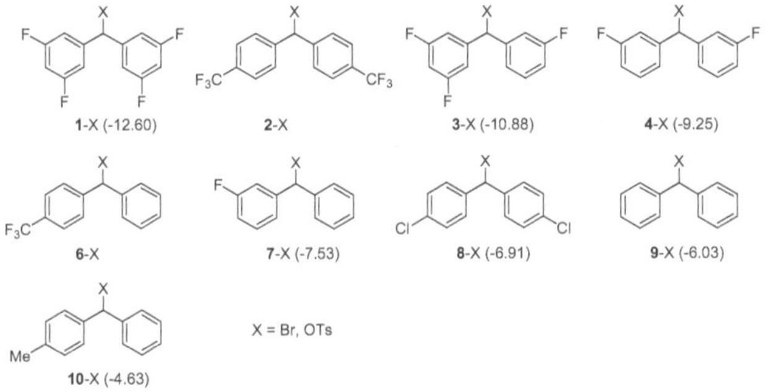

1-X (-12.60)   2-X   3-X (-10.88)   4-X (-9.25)

6-X   7-X (-7.53)   8-X (-6.91)   9-X (-6.03)

10-X (-4.63)   X = Br, OTs

According to the sum of σ parameters[1] and E_f parameters,[2] the tetrakis(*meta*-fluoro)-substituted benzhydrylium ion (**1⁺**) is more electron-deficient than the bis-(trifluoromethyl)-substituted cation **2⁺** used in chapter 3, whereas **3⁺**, **4⁺** and **7⁺** are less stabilized than **8⁺** but better stabilized than **2⁺**.

4.2. Results and Discussion

As described for several benzhydryl bromides in the preceding chapter, dissolution of compounds (**1,3,4,7**)-X (X = Br or OTs) in solutions of amines in DMSO led to a monoexponential increase of the conductance due to the generation of HBr or HOTs, which reacted with excess amine to give the corresponding ammonium salts. The rates of these reactions can be followed by conductometry (conductometers: Tacussel CD 810 or

4. Leaving Group Dependence of the S_N1/S_N2 Ratio

Radiometer Analytical CDM 230, Pt electrode: WTW LTA 1/NS). The temperature of the solutions was kept constant (20.0 ± 0.1 °C) by using a circulating bath thermostat. Fitting the time-dependent increase of conductance to the exponential function 4.1 yielded the observed rate constants k_{obs}

$$dG/dt = G_{max}[1 - e^{(-k_{obs}t)}] + C \qquad (4.1)$$

As in the examples described in Chapter 3, plots of k_{obs} versus the concentrations of the amines were linear (Figure 4.1) and almost went through the origin indicating that the S_N1 mechanism is neglible when amines are present. However, very small ionization rate constants (k_1) for the reactions of benzhydryl bromides have also been measured in the absence of nucleophilic amines. In order to prevent autocatalysis by generated HBr the rates of these reactions have been measured in the presence of 20 eq. of 2,6-lutidine. As expressed by equation 4.2, the observed rate constants k_{obs} can be regarded as the sum of a very small amine-independent term k_1 and an amine-dependent term k_2[amine], which are summarized in Table 4.1 and 4.2.

$$k_{obs} = k_1 + k_2[\text{amine}] \qquad (4.2)$$

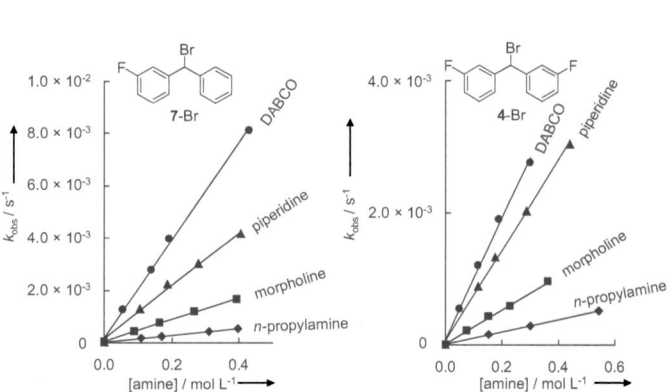

Figure 4.1. Plots of k_{obs} (s^{-1}) of the reactions of different benzhydryl bromides with amines in DMSO at 20 °C vs. the concentrations of the amines.

4. Leaving Group Dependence of the S_N1/S_N2 Ratio

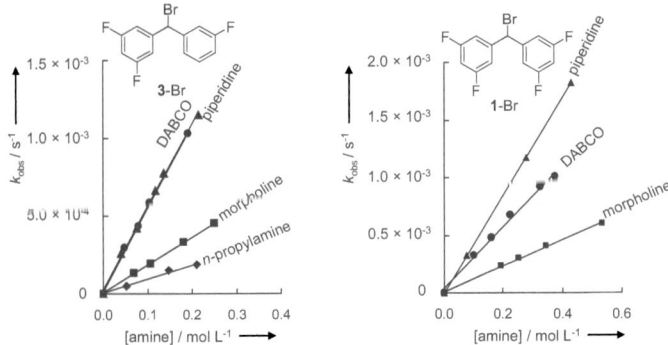

Figure 4.1. (continued)

Table 4.1. Rate constants (at 20° C) for the solvolyses of the benzhydryl bromides in DMSO (s^{-1}) and for their reactions with amines in DMSO ($M^{-1}s^{-1}$).

N°-Br	DMSO (k_1)[a]	DABCO (k_2)	piperidine (k_2)	morpholine (k_2)	n-PrNH$_2$ (k_2)
1-Br	9.62×10^{-7}	2.69×10^{-3}	4.24×10^{-3}	1.12×10^{-3}	-
2-Br[b]	2.76×10^{-6}	-	6.66×10^{-3}	2.17×10^{-3}	1.17×10^{-3}
3-Br	2.84×10^{-6}	5.33×10^{-3}	5.34×10^{-3}	1.79×10^{-3}	8.99×10^{-4}
4-Br	5.92×10^{-6}	9.23×10^{-3}	6.85×10^{-3}	2.57×10^{-3}	9.40×10^{-4}
6-Br[b]	1.25×10^{-5}	-	9.36×10^{-3}	3.29×10^{-3}	1.13×10^{-3}
7-Br	3.37×10^{-5}	1.87×10^{-2}	1.02×10^{-2}	4.18×10^{-3}	1.25×10^{-3}
8-Br[b]	1.36×10^{-4}		2.33×10^{-2}	9.51×10^{-3}	2.19×10^{-3}
9-Br[b]	5.45×10^{-4}	5.45×10^{-2}	1.69×10^{-2}	7.30×10^{-3}	1.33×10^{-3}
10-Br[b]	6.71×10^{-3}	1.92×10^{-1}	3.57×10^{-2}	2.16×10^{-2}	3.98×10^{-3}

[a] determined independently in the presence of 20 eq. of 2,6-lutidine
[b] data taken from Chapter 3

First-order rate constants (k_1) from this and the preceding chapter were used for the Hammett plot in Figure 4.2.

4. Leaving Group Dependence of the S_N1/S_N2 Ratio

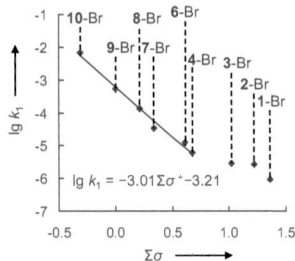

Figure 4.2. Plot of lg k_1 of the solvolysis reactions of the benzhydryl bromides in DMSO vs. Hammett's substituent constants σ^+ (σ_m for **1,3,4,7-Br**)[1]; benzhydryl bromides with a $\Sigma\sigma > 0.68$ were not used for the correlation.

The first-order rate constants (k_1) for the solvolyses of the monofluoro **7-Br** and the difluoro substituted compound **4-Br** in DMSO fit nicely on the correlation line of the first-order rate constants of (**10,9,8,5**)-Br in DMSO reported in chapter 3, yielding an almost unchanged ρ parameter of -3.01 (-2.94 when the rate constants of **7-Br** and **4-Br** are excluded). Therefore, it can be assumed that the rate constants k_1 for **7-Br** and **4-Br** in DMSO also refer to the ionization step. More importantly, a result from the preceding chapter is confirmed. The rate constants of **3-Br** and **1-Br** deviate upward from the correlation line. The tetrakis(*meta*-fluoro)-substituted benzhydryl bromide **1-Br** reacts approximately 20 times faster than expected for an S_N1 reaction (Figure 4.3).

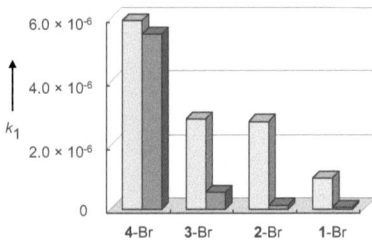

Figure 4.3. Rate constants k_1 of the solvolysis reactions of benzhydryl bromides **1-4-Br** in DMSO (at 20 °C) (in green) compared with calculated rate constants (in red) derived from the Hammett plot (Figure 4.2) $k_1 = 10^{(-3.01\sigma + -3.21)}$.

174

4. Leaving Group Dependence of the S_N1/S_N2 Ratio

Thus, the S_N2-type reaction with DMSO, previously encountered for the reaction of **2-Br** (Chapter 3), is obviously a general behavior of acceptor substituted benzhydryl bromides. Since the lifetimes for the carbocations generated from **3-Br** and **1-Br** are expected to be considerably shorter than 1×10^{-14} s, this observation is also in line with the previously stated lifetime argument.

The amine-dependent rate constants k_2 for the reactions of (**1,3,4,7**)-Br were included in the Hammett plot from the preceding chapter (Figure 3.3). The resulting Hammett plot (Figure 4.4) shows an almost unchanged situation. The rate constants k_2 decrease only by factors of 8 to 71 when going from the mono methyl-substituted benzhydryl bromide **10-Br** to the tetrafluoro-substituted compound **1-Br** (Table 4.1), whereas the rate constant for the amine independent first order rate constant k_1 decreases by a factor of 6975 (Table 4.1, Figure 4.2).

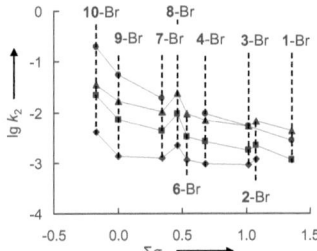

Figure 4.4. Plot of lg k_2 of the reactions of the benzhydryl bromides with amines in DMSO at 20 °C vs. Hammett´s substituent constants σ.[1] ● for DABCO ▲ for piperidine ■ for morpholine ♦ for n-propylamine.

When only the rate constants for the fluorinated benzhydryl bromides are considered, a good linear Hammett correlation is obtained with ρ ranging from -0.21 to -0.81 (Figure 4.5). Probably, the high quality of this correlation is due to the fact the only meta substituents are considered which operate exclusively by the inductive effect.

4. Leaving Group Dependence of the S_N1/S_N2 Ratio

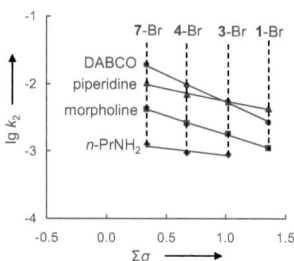

Figure 4.5. Plot of lg k_2 of the reactions of the benzhydryl bromides with amines in DMSO at 20 °C vs. Hammett's substituent constants σ.[1] (Data from the preceding chapter was excluded) ● for DABCO ($\rho = -0.81\sigma - 1.46$) ▲ for piperidine ($\rho = -0.37\sigma - 1.89$) ■ for morpholine ($\rho = -0.55\sigma - 2.20$) ♦ for n-propylamine ($\rho = -0.21\sigma - 2.85$).

As depicted in Figure 4.5, the most negative reaction constant ($\rho = -0.81$) is observed for the highly nucleophilic DABCO and the smallest sensitivity ($\rho = -0.21$) is observed for n-propylamine with the lowest nucleophilicity in this series. As stated in Chapter 3, these values suggest that the S_N2 reactions proceed via transition states where little positive charge is developed at the benzhydryl center. In order to investigate the influence of the leaving group on the S_N1/S_N2 ratio, the reactions of benzhydryl tosylates (**7,4,3,1**)-OTs with amines in DMSO were investigated using the same methodology. As for the benzhydryl bromides, plots of the observed rate constants (k_{obs}) versus the concentration of amine were linear according to equation 4.2 (Figure 4.6).

4. Leaving Group Dependence of the S_N1/S_N2 Ratio

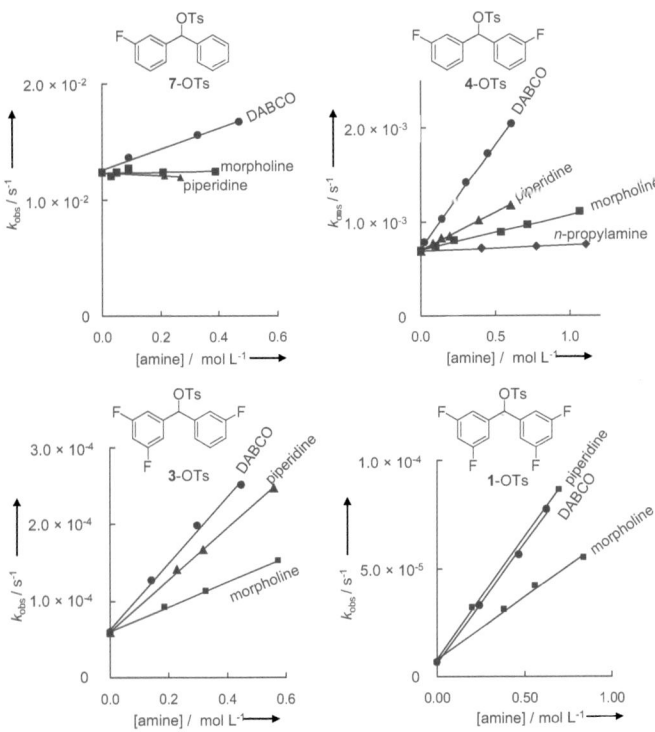

Figure 4.6. Plots of k_{obs} (s^{-1}) of the reactions of different benzhydryl tosylates with amines in DMSO at 20 °C vs. the concentrations of the amines.

Amine-independent rate constants k_1 and amine-dependent rate constants k_2 are summarized in Table 4.2.

4. Leaving Group Dependence of the S_N1/S_N2 Ratio

Table 4.2. Rate constants (at 20° C) for the solvolyses of the benzhydryl tosylates in DMSO (s^{-1}) and for their reactions with amines in DMSO ($M^{-1}s^{-1}$).

nucleophiles	7-OTs	4-OTs	3-OTs	1-OTs
DMSO (k_1)	1.23×10^{-2}	6.87×10^{-4}	5.85×10^{-5}	6.76×10^{-6}
DABCO (k_2)	8.90×10^{-3}	2.22×10^{-3}	4.31×10^{-4}	1.11×10^{-4}
piperidine (k_2)	-	7.97×10^{-4}	3.38×10^{-4}	1.13×10^{-4}
morpholine (k_2)	-	3.77×10^{-4}	1.62×10^{-4}	5.80×10^{-5}
n-PrNH$_2$ (k_2)	-	6.26×10^{-5}	8.48×10^{-5}	5.77×10^{-5}

As shown in Figure 4.6 and in Table 4.2, the solvolysis reactions (k_1) of the benzhydryl tosylates (**7,4,3,1**)-OTs are 7 to 366 times faster in DMSO than the solvolyses of benzhydryl bromides (**7,4,3,1**)-Br. The small ratio of 7 between the solvolysis rate constants k_1 for bromide and tosylate is observed for the tetrafluoro-substituted benzhydryl derivatives **1-X**. This small ratio can be explained by a direct nucleophilic attack of DMSO (S_N2 reaction with DMSO) on the benzhydryl bromides (**1,2,3**)-Br, which is not observed for the benzhydryl tosylates (**7,4,3,1**)-OTs despite the highly destabilized character of the corresponding carbocations.

A Hammett plot of the solvolysis rate constants k_1 for the benzhydryl tosylates (**7,4,3,1**)-OTs (Figure 4.7) showed a linear correlation with a slope of $\rho = -3.19$. A similar value of the reaction constant ($\rho = -3.01$) has been found for the solvolyses of benzhydryl bromides which ionize without nucleophilic assistance by DMSO (Figure 4.2). Therefore, the first-order rate constants k_1 for the reaction of benzhydryl tosylates **1-4-OTs** may be assigned to ionization processes without nucleophilic assistance from the solvent DMSO. However, this conclusion is questioned by the observation that the ratio k_{ROH}/k_{DMSO} for benzhydryl tosylates decreases noticeably with decreasing electrofugality of the benzhydrylium ions (see below).

4. Leaving Group Dependence of the S_N1/S_N2 Ratio

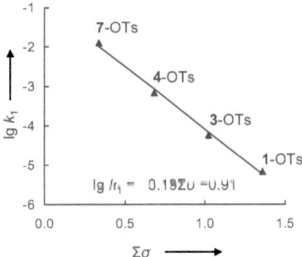

Figure 4.7. Plot of lg k_1 for the solvolysis reactions of (**7,4,3,1**)-OTs in DMSO at 20 °C vs. Hammett's substituent constants σ_m.[1]

Since the amine-independent rate constants k_1 for benzhydryl bromides with $E_f \geq -9.26$ and for all benzhydryl tosylates investigated correspond to the ionization reaction, they can be employed to derive nucleofugality parameters N_f and s_f for bromide and tosylate in neat DMSO.[2] Plots of lg k_1 versus the electrofugality parameters E_f resulted in linear correlations according to equation 4.3 (Figure 4.8).

$$\lg k_1 = s_f(N_f + E_f) \qquad (4.3)$$

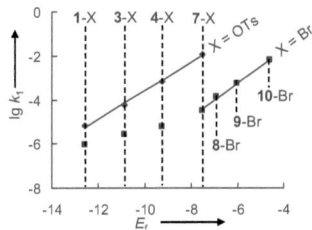

Figure 4.8. Plot of lg k_1 for the solvolysis of various substituted benzhydryl bromides and tosylates in DMSO at 20 °C vs. electrofugality E_f.

From these correlations, one can extract the nucleofugality parameters N_f as the negative intercepts on the abscissa (E_f axis) and the s_f parameters as the slopes of the correlation lines (Table 4.3). Due to the S_N2 type reaction of electron-poor benzhydryl bromides, rate

4. Leaving Group Dependence of the S_N1/S_N2 Ratio

constants $k_1 < 3 \times 10^{-5}$ s^{-1}, for the ionization of benzhydryl bromides, were excluded from the correlation line and the calculation of nucleofugality parameters of bromide in DMSO.

Table 4.3. Nucleofugality parameters N_f and s_f for bromide and tosylate at 20 °C in DMSO. Parameters for chloride at 25 °C taken from Ref. [3]

X	N_f / s_f
OTs	4.45 / 0.64
Br	1.84 / 0.78
Cl	0.35 / 1.30

These nucleofugality parameters allow a direct comparison of the leaving group abilities of chloride, bromide and tosylate in DMSO. The slope parameters s_N for OTs and Br are lower than their slope parameters in most protic solvents, with the consequence that the reactivity ratio k_{ROH}/k_{DMSO} of benzhydryl tosylates decreases with decreasing electrofugality of the benzhydrylium ions as illustrated for the solvolyses of benzhydrylium tosylates in DMSO, EtOH, and MeOH in Figure 4.9. Obviously, destabilization of the benzhydrylium ions by electron acceptor substituents affects the solvation by DMSO more than the solvation by alcohols, but the nature of this solvation-nucleophilic solvent participation or nucleophilic solvation is not clear.[4] The decreasing k_{OTs}/k_{Br} ratio with decreasing electrofugality of the benzhydrylium ions (Figure 4.9) furthermore shows that nucleophilic solvent participation in DMSO is much more important in solvolyses of benzhydryl bromides than of benzhydryl tosylates.

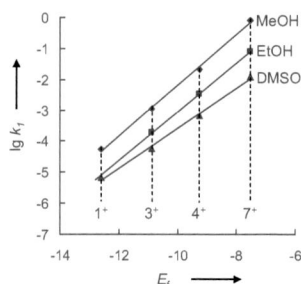

4. Leaving Group Dependence of the S_N1/S_N2 Ratio

Figure 4.9. Plot of lg k_1 for the solvolysis of benzhydryl tosylates (**1,3,4,7**)-OTs in methanol, ethanol at 25 °C and DMSO at 20 °C vs. electrofugality E_f.

Like in protic solvents, tosylate is also the best leaving group in DMSO. Thus, the benzhydryl tosylate **7**-OTs solvolyses in DMSO at 20 °C with a half-life of 56 s, the half-life for **7**-Br is 6 h (derived from measured rate constants), and for the benzhydryl chloride **7**-Cl a half-life of 47 years at 25 °C is calculated from the nucleofugality parameters presented in Table 4.3.

The amine-dependent second-order rate constants k_2 (Table 4.2) show that the reactions of benzhydryl tosylates (**7,4,3,1**)-OTs with nucleophilic amines are 2 to 37 times slower than the corresponding reactions with benzhydryl bromides, showing that the relative leaving group abilities of TsO⁻ and Br⁻ are opposite in S_N1 and S_N2 reactions. Hammett plots of the amine-dependent second-order rate constant k_2 vs. $\Sigma\sigma$ (Figure 4.10) resulted in linear correlations and yielded reaction constants ρ ranging from -1.89 to -0.05. The most negative reaction constant ρ was observed for the reactions with DABCO and a reaction constant ρ of almost zero was observed for the reactions with n-propylamine. While the dependence of the reaction constants ρ on the nature of the amines show exactly the same trends in Figure 4.10 and 4.5, the absolute value of ρ is more than two times larger for reactions of DABCO with benzhydryl tosylates than with benzhydryl bromides.

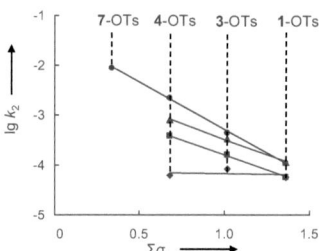

Figure 4.10. Plot of lg k_2 of the reactions of the benzhydryl tosylates with amines vs. Hammett´s substituent constants σ.[1] (Data from the preceding chapter was excluded) ● for DABCO ▲ for piperidine ■ for morpholine ♦ for n-propylamine.

4. Leaving Group Dependence of the S_N1/S_N2 Ratio

Reaction constants $\rho \approx 0$ for the reactions of benzhydryl bromides and tosylates with n-propylamine (Figures 4.5 and 4.10) show that in these series the charge density at the benzhydryl carbon is not charged from reactants to the transition state. The negative reaction constants for the reaction with DABCO indicate a more carbocationic transition state, i.e., a stronger imbalance between bond formation and bond breaking for the reaction with DABCO.[5] The observation that the reactions with more nucleophilic DABCO have more negative reaction constants is surprising because Tsuno et al. observed the opposite trend when investigating the reactions of benzyl tosylates with substituted N,N-dimethylanilines.[6] In their investigations the S_N2 reactions of donor substituted (and thus more nucleophilic) N,N-dimethylanilines were only weakly affected by substituents on the benzyl tosylate, while the S_N2 reactions of acceptor substituted N,N-dimethylanilines were highly influenced by the benzyl substitution pattern.

4.3. Conclusion and Outlook

Though the order of nucleofugalities OTs > Br > Cl in ionization processes is the same in DMSO as in most protic solvents, there are significant differences of the relative reactivities in these three solvents as illustrated for the parent benzhydryl derivatives **9-X** in Figure 4.11. In TFE, the unsubstituted benzhydryl bromide (**9-Br**) solvolyses approximately equally fast as the benzhydryl chloride. The k_{Br}/k_{Cl} ratio increases to a factor of 15 to 45 in aqueous ethanol, ethanol and aqueous acetone and adopts a value of 1.3×10^4 in DMSO.

4. Leaving Group Dependence of the S_N1/S_N2 Ratio

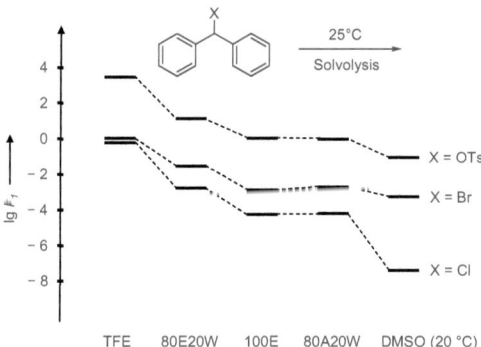

Figure 4.11. Comparison of calculated lg k_1 for the solvolyses reactions of benzhydryl derivatives in a series of solvents.

On the other hand, the almost constant k_{OTs}/k_{Br} ratio (2.1 × 10³ to 4.9 × 10²) in all protic solvents shown in Figure 4.10 shrinks to a value of 1.7 × 10² in DMSO. As illustrated in Figure 4.12, the solvolysis rate constants correlate well with the solvent ionizing power Y_{xBnBr} of protic solvents derived from secondary benzyl bromides.[3,7] As k_{OTs}/k_{Br} is smaller in DMSO and k_{Br}/k_{Cl} is larger in DMSO than in protic solvents it its not possible to define a Y_{xBnBr} value which holds for different leaving groups.

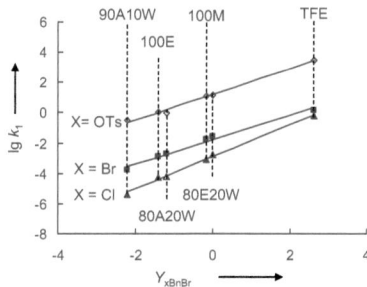

Figure 4.12 Plot of lg k_1 for the reactions of the benzhydryl derivatives Ph₂CHX (**9-X**) vs. the solvent ionizing power Y_{xBnBr}.[6] Data point without filling are calculated solvolysis rate constants according to equation 4.3. Data points with filling are measured rate constants taken from Ref. [2].

183

4. Leaving Group Dependence of the S_N1/S_N2 Ratio

A moderate correlation is also achieved, when lg k_1 is plotted vs. the empirical solvent polarity parameter E_T^N (Figure 4.11).[8] Comparable observations where reported by Dvorko et al., for the heterolysis of *tert*-butyl chloride in a series of solvents.[9,10]

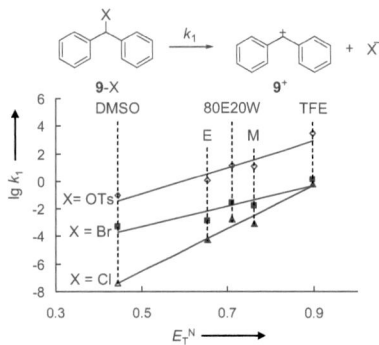

Figure 4.11 Plot of lg k_1 for the reactions of the benzhydryl derivatives (Ph$_2$CHX) vs. the solvent polarity parameter E_T^N. Data point without filling are calculated solvolysis rate constants according to equation 4.3. Data points with filling are measured rate constants taken from Ref. [2].

Unequivocal evidence for nucleophilic solvent participation was so far only observed for the solvolyses of the acceptor substituted benzhydryl bromides (**3-Br**, **2-Br** and **1-Br**) in DMSO. In all other solvents under investigation, nucleophilic solvent participation was not observed, possibly because the narrow reactivity range experimentally accessible.

The second order rate constants k_2 are 2 to 37 times larger for benzhydryl bromides than for benzhydryl tosylates (Table 4.4).

4. Leaving Group Dependence of the S_N1/S_N2 Ratio

Table 4.4. Comparison of second-order rate constants ($k_2/M^{-1}s^{-1}$) for the reaction of benzhydryl bromides and tosylates in DMSO with various amines.

N°-X	amine	k_2 (X=Br)	k_2 (X=OTs)	k_2 (X=Br)/k_2 (X=OTs)
7-X	DABCO	1.87×10^{-2}	8.90×10^{-3}	2.1
	piperidine	1.02×10^{-2}	-	-
	morpholine	4.18×10^{-3}	-	-
	n-PrNH$_2$	1.25×10^{-3}	-	-
4-X	DABCO	9.23×10^{-3}	2.22×10^{-3}	4.2
	piperidine	6.85×10^{-3}	7.97×10^{-4}	8.6
	morpholine	2.57×10^{-3}	3.77×10^{-4}	6.8
	n-PrNH$_2$	9.40×10^{-4}	6.26×10^{-5}	15.0
3-X	DABCO	5.33×10^{-3}	4.31×10^{-4}	12.4
	piperidine	5.34×10^{-3}	3.38×10^{-4}	15.8
	morpholine	1.79×10^{-3}	1.62×10^{-4}	11.0
	n-PrNH$_2$	8.99×10^{-4}	8.48×10^{-5}	10.6
1-X	DABCO	2.69×10^{-3}	1.11×10^{-4}	24.2
	piperidine	4.24×10^{-3}	1.13×10^{-4}	37.5
	morpholine	1.12×10^{-3}	5.80×10^{-5}	19.3
	n-PrNH$_2$	-	5.77×10^{-5}	-

The analysis by Hoz et al. offers a possible explanation for this decrease. According to G2-calculations the barriers for the identity S_N2 reactions,[11] increased significantly when going from right to left in the periodic table, e.g., from halides to chalcogenides (Scheme 4.2).

Scheme 4.2. S_N2 identity reaction investigated by Hoz et al. and calculated intrinsic barrier heights G2(+).

$$X^- + Me-X \longrightarrow Me-X + X^-$$

G2(+) = 55.2 kJ /mol for X = Cl
G2(+) = 45.2 kJ /mol for X = Br
G2(+) = 81.6 kJ /mol for X = OMe
G2(+) = 91.6 kJ /mol for X = SMe

The surprisingly low reactivities of tosylates in S_N2 reactions may, therefore, be due to the higher intrinsic barrier for chalcogenide exchange reactions than for halide exchange

4. Leaving Group Dependence of the S_N1/S_N2 Ratio

reactions (Scheme 4.2).[11] Previously equation 4.4 has been successfully employed to calculate rate constants for S_N2 reactions.[12] In this correlation, nucleophiles are characterized by the nucleophilicity parameter N and the sensitivity parameter s_N while electrophiles are characterized by the electrophilicity parameter E and the sensitivity parameter s_E. Equation 4.4 is an extension of the 1994 Patz equation,[13] and N and s_N are the nucleophile-specific parameters, which have previously been derived from reactions with benzhydrylium ions.

$$\lg k = s_E\, s_N\, (E+N) \qquad (4.4)$$

When plotting ($\lg k_2$)/s_N for the reactions of benzhydryl bromides or tosylates against the nucleophilicity parameters of amines moderate correlations were obtained (Figure 4.12). Because of the relatively small reactivity range covered by Figure 4.12 it is hard to decide whether equation 4.4 is suitable to predict second-order rate constants for these reactions. Similar experiments with benzyl halides are currently under investigation and might help to clarify this situation.

4. Leaving Group Dependence of the S_N1/S_N2 Ratio

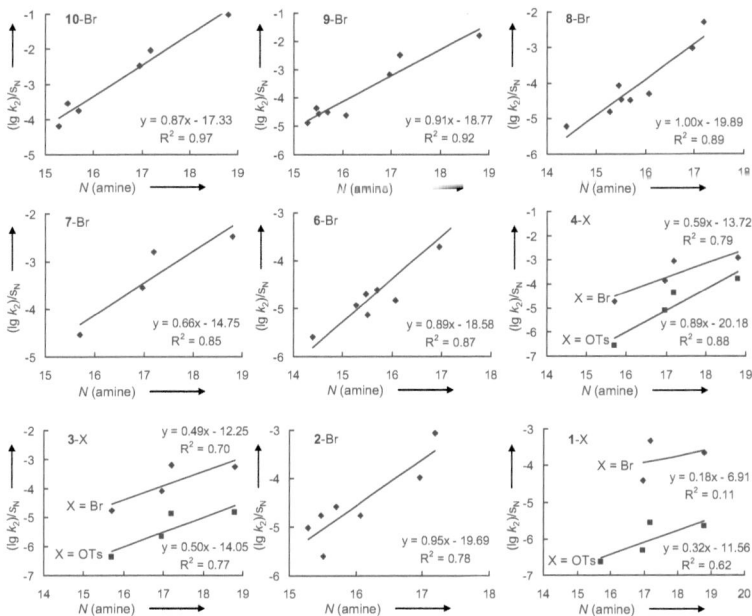

Figure 4.12. Plot of (lg k_2)/s_N for the reactions of the benzhydryl derivatives (Ph$_2$CHX) in DMSO with amines vs. the nucleophilicity parameter N of the amine; amine (N / s_N): DABCO (18.80 / 0.70), piperidine (17.19 / 0.71), morpholine (16.96 / 0.67), ethanolamine (16.07 / 0.61), 1-aminopropan-2-ol (15.47 / 0.65), n-PrNH$_2$ (15.70 / 0.64), benzylamine (15.28 / 0.65), diethanolamine (15.51 / 0.70), 2-aminobutan-1-ol (14.39 / 0.67).

4.4. Experimental section

Method

The methodology described in chapter 3 was analogously used to determine first-order rate constants k_{obs} for the reactions of benzhydryl bromides and tosylates in DMSO. Amine-independent rate constants (k_1) were determined in the presence of 20 eq. 2,6-lutidine.

4. Leaving Group Dependence of the S_N1/S_N2 Ratio

3-Fluorbenzhydryl bromide (7-Br)
20 °C, in DMSO, conductometry

Table 4.5. Individual rate constants for the reaction of **7-Br** in DMSO in the presence of various amines

Nu	[7-Br]$_o$/M	[Nu]/M	k_{obs}/s^{-1}	k_{obs} vs. [Nu] correlation	k_2/M^{-1} s^{-1}
DABCO	2.76 × 10^{-3}	0.00	3.37 × 10^{-4}	y = 1.87E-02x + 1.76E-04; R^2 = 9.98E-01	**1.87 × 10^{-2}**
	2.77 × 10^{-3}	0.06	1.27 × 10^{-3}		
	3.00 × 10^{-3}	0.14	2.78 × 10^{-3}		
	3.45 × 10^{-3}	0.19	3.98 × 10^{-3}		
	3.16 × 10^{-3}	0.43	8.13 × 10^{-3}		
piperidine	2.76 × 10^{-3}	0.00	3.37 × 10^{-4}	y = 1.02E-02x + 1.69E-04; R^2 = 9.94E-01	**1.02 × 10^{-2}**
	2.76 × 10^{-3}	0.11	1.72 × 10^{-4}		
	2.72 × 10^{-3}	0.19	2.60 × 10^{-4}		
	2.77 × 10^{-3}	0.28	4.22 × 10^{-4}		
	2.75 × 10^{-3}	0.40	5.34 × 10^{-4}		
morpholine	2.76 × 10^{-3}	0.00	3.37 × 10^{-4}	y = 4.18E-03x + 5.88E-05; R^2 = 9.98E-01	**4.18 × 10^{-3}**
	2.87 × 10^{-3}	0.09	4.38 × 10^{-4}		
	2.89 × 10^{-3}	0.16	7.65 × 10^{-4}		
	2.81 × 10^{-3}	0.27	1.20 × 10^{-3}		
	2.70 × 10^{-3}	0.39	1.68 × 10^{-3}		

4. Leaving Group Dependence of the S_N1/S_N2 Ratio

Table 4.5. (continued)

Nu	$[7\text{-Br}]_o$/M	[Nu]/M	k_{obs}/s^{-1}	k_{obs} vs. [Nu] correlation	$k_2/M^{-1} s^{-1}$
n-PrNH$_2$	2.76×10^{-3}	0.00	3.37×10^{-4}	$y = 1.25\text{E-}03x + 3.75\text{E-}05$; $R^2 = 9.99\text{E-}01$	1.25×10^{-3}
	2.87×10^{-3}	0.11	1.72×10^{-4}		
	2.85×10^{-3}	0.17	2.60×10^{-4}		
	2.00×10^{-3}	0.31	4.22×10^{-4}		
	2.86×10^{-3}	0.40	5.34×10^{-4}		

3,3'-Difluorbenzhydryl bromide (4-Br)
20 °C, in DMSO, conductometry

Table 4.6. Individual rate constants for the reaction of **4-Br** in DMSO in the presence of various amines

Nu	$[4\text{-Br}]_o$ M	[Nu]/M	k_{obs}/s^{-1}	k_{obs} vs. [Nu] correlation	$k_2/M^{-1} s^{-1}$
DABCO	2.96×10^{-3}	0.00	5.92×10^{-6}	$y = 9.23\text{E-}03x + 6.65\text{E-}05$; $R^2 = 9.96\text{E-}01$	9.23×10^{-3}
	2.45×10^{-3}	0.05	5.38×10^{-4}		
	2.82×10^{-3}	0.12	1.20×10^{-3}		
	3.07×10^{-3}	0.19	1.90×10^{-3}		
	2.84×10^{-3}	0.30	2.76×10^{-3}		
piperidine	2.96×10^{-3}	0.00	5.92×10^{-6}	$y = 6.85\text{E-}03x + 6.46\text{E-}05$; $R^2 = 9.98\text{E-}01$	6.85×10^{-3}
	2.98×10^{-3}	0.12	8.83×10^{-4}		
	2.98×10^{-3}	0.18	1.33×10^{-3}		
	2.88×10^{-3}	0.29	2.03×10^{-3}		
	2.87×10^{-3}	0.44	3.05×10^{-3}		
morpholine	2.96×10^{-3}	0.00	5.92×10^{-6}	$y = 2.57\text{E-}03x + 1.34\text{E-}05$; $R^2 = 9.98\text{E-}01$	2.57×10^{-3}
	3.06×10^{-3}	0.08	2.11×10^{-4}		
	2.95×10^{-3}	0.15	4.33×10^{-4}		
	2.95×10^{-3}	0.23	5.83×10^{-4}		
	2.74×10^{-3}	0.36	9.51×10^{-4}		

4. Leaving Group Dependence of the S_N1/S_N2 Ratio

Table 4.6. (continued)

Nu	$[4\text{-Br}]_o/M$	$[Nu]/M$	k_{obs}/s^{-1}	k_{obs} vs. [Nu] correlation	$k_2/M^{-1}\,s^{-1}$
n-PrNH$_2$	2.96×10^{-3}	0.00	5.92×10^{-6}	$y = 9.40\text{E-}04x + 9.18\text{E-}06$; $R^2 = 1.00\text{E+}00$	$\mathbf{9.40 \times 10^{-4}}$
	2.55×10^{-3}	0.15	1.60×10^{-4}		
	2.84×10^{-3}	0.30	2.88×10^{-4}		
	2.88×10^{-3}	0.54	5.19×10^{-4}		

3,3',5-Trifluorbenzhydryl bromide (3-Br)
20 °C, in DMSO, conductometry

Table 4.7. Individual rate constants for the reaction of **3-Br** in DMSO in the presence of various amines

Nu	$[3\text{-Br}]_o/M$	$[Nu]/M$	k_{obs}/s^{-1}	k_{obs} vs. [Nu] correlation	$k_2/M^{-1}\,s^{-1}$
DABCO	2.43×10^{-3}	0.00	2.84×10^{-6}	$y = 5.33\text{E-}03x + 1.51\text{E-}05$; $R^2 = 9.99\text{E-}01$	$\mathbf{5.33 \times 10^{-3}}$
	2.34×10^{-3}	0.05	2.97×10^{-4}		
	2.64×10^{-3}	0.08	4.35×10^{-4}		
	2.44×10^{-3}	0.11	5.84×10^{-4}		
	2.28×10^{-3}	0.19	1.03×10^{-3}		
piperidine	2.43×10^{-3}	0.00	2.84×10^{-6}	$y = 5.34\text{E-}03x + 2.62\text{E-}05$; $R^2 = 9.98\text{E-}01$	$\mathbf{5.34 \times 10^{-3}}$
	2.39×10^{-3}	0.04	2.61×10^{-4}		
	2.47×10^{-3}	0.08	4.23×10^{-4}		
	2.68×10^{-3}	0.12	6.65×10^{-4}		
	2.33×10^{-3}	0.14	7.76×10^{-4}		
	2.51×10^{-3}	0.21	1.15×10^{-3}		
morpholine	2.96×10^{-3}	0.00	2.84×10^{-6}	$y = 1.79\text{E-}03x + 6.15\text{E-}06$; $R^2 = 1.00\text{E+}00$	$\mathbf{1.79 \times 10^{-3}}$
	2.64×10^{-3}	0.07	1.35×10^{-4}		
	2.63×10^{-3}	0.11	1.95×10^{-4}		
	2.96×10^{-3}	0.18	3.29×10^{-4}		
	2.74×10^{-3}	0.25	4.53×10^{-4}		

4. Leaving Group Dependence of the S_N1/S_N2 Ratio

Table 4.7. (continued)

Nu	[3-Br]$_o$/M	[Nu]/M	k_{obs}/s^{-1}	k_{obs} vs. [Nu] correlation	k_2/M^{-1} s^{-1}
n-PrNH$_2$	2.96 × 10^{-3}	0.00	2.84 × 10^{-6}	y = 8.99E-04x + 4.19E-06; R^2 = 9.90E-01	**8.99 × 10^{-4}**
	2.67 × 10^{-3}	0.05	4.76 × 10^{-5}		
	2.62 × 10^{-3}	0.15	1.49 × 10^{-4}		
	2.66 × 10^{-3}	0.21	1.86 × 10^{-4}		

3,3',5,5'-Tetrafluorbenzhydryl bromide (1-Br)
20 °C, in DMSO, conductometry

Table 4.8. Individual rate constants for the reaction of 1-Br in DMSO in the presence of various amines

Nu	[1-Br]$_o$/M	[Nu]/M	k_{obs}/s^{-1}	k_{obs} vs. [Nu] correlation	k_2/M^{-1} s^{-1}
DABCO	2.23 × 10^{-3}	0.00	9.62 × 10^{-7} *	y = 2.69E-03x + 3.56E-05; R^2 = 9.94E-01	**2.69 × 10^{-3}**
	2.42 × 10^{-3}	0.10	3.26 × 10^{-4}		
	2.65 × 10^{-3}	0.16	4.79 × 10^{-4}		
	2.93 × 10^{-3}	0.23	6.78 × 10^{-4}		
	1.39 × 10^{-3}	0.33	9.17 × 10^{-4}		
	2.51 × 10^{-3}	0.37	1.01 × 10^{-3}		
piperidine	2.43 × 10^{-3}	0.00	2.84 × 10^{-6} *	y = 4.24E-03x + 6.20E-07; R^2 = 1.00E+00	**4.24 × 10^{-3}**
	2.71 × 10^{-3}	0.08	3.25 × 10^{-4}		
	2.80 × 10^{-3}	0.28	1.17 × 10^{-3}		
	2.61 × 10^{-3}	0.43	1.82 × 10^{-3}		

4. Leaving Group Dependence of the S_N1/S_N2 Ratio

Table 4.8. (continued)

Nu	$[1\text{-Br}]_o$/M	[Nu]/M	k_{obs}/s^{-1}	k_{obs} vs. [Nu] correlation	$k_2/M^{-1}\,s^{-1}$
morpholine	2.23×10^{-3}	0.00	9.62×10^{-7} *	y = 1.12E-03x + 1.12E-05; R^2 = 9.98E-01	1.12×10^{-3}
	1.39×10^{-3}	0.19	2.33×10^{-4}		
	1.45×10^{-3}	0.25	3.01×10^{-4}		
	1.64×10^{-3}	0.34	4.06×10^{-4}		
	1.51×10^{-3}	0.53	6.02×10^{-4}		

* reaction was only followed for about 0.4 half-times (2.6×10^5 s = 3 d).

3-Fluorbenzhydryl tosylate (7-OTs)

20 °C, in DMSO, conductometry

Table 4.9. Rate constants for the reaction of **7-OTs** in DMSO in the presence of various amines

Nu	$[\text{7-OTs}]_o$/M	[Nu]/M	k_{obs}/s^{-1}	k_{obs} vs. [Nu] correlation	$k_2/M^{-1}\,s^{-1}$
DABCO	1.74×10^{-3}	0.00	1.23×10^{-2}	y = 8.90E-03x + 1.26E-02; R^2 = 9.86E-01	8.90×10^{-3}
	1.62×10^{-3}	0.09	1.37×10^{-2}		
	1.65×10^{-3}	0.33	1.55×10^{-2}		
	1.62×10^{-3}	0.47	1.67×10^{-2}		
piperidine	1.74×10^{-3}	0.00	1.23×10^{-2}	y = -1.28E-03x + 1.23E-02; R^2 = 9.06E-01	–
	1.58×10^{-3}	0.21	1.21×10^{-2}		
	1.65×10^{-3}	0.27	1.19×10^{-2}		
morpholine	1.74×10^{-3}	0.00	1.23×10^{-2}	y = 4.76E-04x + 1.23E-02; R^2 = 1.14E-01	–
	2.36×10^{-3}	0.03	1.20×10^{-2}		
	2.37×10^{-3}	0.05	1.23×10^{-2}		
	2.36×10^{-3}	0.09	1.26×10^{-2}		
	2.27×10^{-3}	0.21	1.23×10^{-2}		
	2.21×10^{-3}	0.39	1.24×10^{-2}		

4. Leaving Group Dependence of the S_N1/S_N2 Ratio

3,3'-Difluorbenzhydryl tosylate (4-OTs)
20 °C, in DMSO, conductometry

Table 4.10. Individual rate constants for the reaction of **4-OTs** in DMSO in the presence of various amines

Nu	[4-OTs]$_o$/M	[Nu]/M	k_{obs}/s^{-1}	k_{obs} vs. [Nu] correlation	k_2/M^{-1} s^{-1}
DABCO	2.54 × 10^{-3}	0.00	6.87 × 10^{-4}	$y = 2.22\text{E-03}x + 7.08\text{E-04}$; $R^2 = 9.99\text{E-01}$	2.22 × 10^{-3}
	2.67 × 10^{-3}	0.03	7.78 × 10^{-4}		
	2.92 × 10^{-3}	0.14	1.02 × 10^{-3}		
	3.06 × 10^{-3}	0.31	1.41 × 10^{-3}		
	2.69 × 10^{-3}	0.45	1.72 × 10^{-3}		
	2.82 × 10^{-3}	0.61	2.04 × 10^{-3}		
piperidine	2.54 × 10^{-3}	0.00	6.87 × 10^{-4}	$y = 7.97\text{E-04}x + 7.01\text{E-04}$; $R^2 = 9.96\text{E-01}$	7.97 × 10^{-4}
	2.46 × 10^{-3}	0.08	7.67 × 10^{-4}		
	2.01 × 10^{-3}	0.14	8.29 × 10^{-4}		
	1.90 × 10^{-3}	0.20	8.50 × 10^{-4}		
	1.92 × 10^{-3}	0.39	1.02 × 10^{-3}		
	1.92 × 10^{-3}	0.61	1.18 × 10^{-3}		
morpholine	2.54 × 10^{-3}	0.00	6.87 × 10^{-4}	$y = 3.77\text{E-04}x + 6.97\text{E-04}$; $R^2 = 9.97\text{E-01}$	3.77 × 10^{-4}
	2.92 × 10^{-3}	0.10	7.42 × 10^{-4}		
	2.13 × 10^{-3}	0.23	7.92 × 10^{-4}		
	3.01 × 10^{-3}	0.54	8.91 × 10^{-4}		
	3.05 × 10^{-3}	0.72	9.67 × 10^{-4}		
	2.84 × 10^{-3}	1.07	1.10 × 10^{-4}		
n-PrNH$_2$	2.54 × 10^{-3}	0.00	6.87 × 10^{-4}	$y = 6.26\text{E-05}x + 6.88\text{E-04}$; $R^2 = 9.85\text{E-01}$	6.26 × 10^{-5}
	2.00 × 10^{-3}	0.41	7.13 × 10^{-4}		
	1.97 × 10^{-3}	0.77	7.42 × 10^{-4}		
	2.02 × 10^{-3}	1.11	7.54 × 10^{-4}		

4. Leaving Group Dependence of the S_N1/S_N2 Ratio

3,3',5-Trifluorbenzhydryl tosylate (3-OTs)
20 °C, in DMSO, conductometry

Table 4.11. Individual rate constants for the reaction of **3-OTs** in DMSO in the presence of various amines

Nu	[3-OTs]$_o$/M	[Nu]/M	k_{obs}/s^{-1}	k_{obs} vs. [Nu] correlation	$k_2/M^{-1} s^{-1}$
DABCO	2.16 × 10^{-3}	0.00	5.85 × 10^{-5}	$y = 4.31\text{E-}04x + 6.30\text{E-}05$; $R^2 = 9.96\text{E-}01$	4.31 × 10^{-4}
	2.67 × 10^{-3}	0.14	1.27 × 10^{-4}		
	2.92 × 10^{-3}	0.30	1.98 × 10^{-4}		
	3.00 × 10^{-3}	0.45	2.51 × 10^{-4}		
piperidine	2.16 × 10^{-3}	0.00	5.85 × 10^{-5}	$y = 3.38\text{E-}04x + 6.06\text{E-}05$; $R^2 = 9.99\text{E-}01$	3.38 × 10^{-4}
	2.75 × 10^{-3}	0.23	1.42 × 10^{-4}		
	2.34 × 10^{-3}	0.32	167 × 10^{-4}		
	2.45 × 10^{-3}	0.56	2.48 × 10^{-4}		
morpholine	2.16 × 10^{-3}	0.00	5.85 × 10^{-5}	$y = 1.62\text{E-}04x + 5.98\text{E-}05$; $R^2 = 9.99\text{E-}01$	1.62 × 10^{-4}
	2.68 × 10^{-3}	0.19	9.20 × 10^{-5}		
	2.36 × 10^{-3}	0.33	1.13 × 10^{-4}		
	2.62 × 10^{-3}	0.57	1.52 × 10^{-4}		
n-PrNH$_2$	2.16 × 10^{-3}	0.00	5.85 × 10^{-5}	$y = 8.48\text{E-}05x + 5.88\text{E-}05$; $R^2 = 9.99\text{E-}01$	8.48 × 10^{-5}
	2.72 × 10^{-3}	0.39	9.30 × 10^{-5}		
	2.55 × 10^{-3}	0.60	1.09 × 1^{-4}		

4. Leaving Group Dependence of the S_N1/S_N2 Ratio

3,3',5,5'-Tetrafluorbenzhydryl tosylate (1-OTs)
20 °C, in DMSO, conductometry

Table 4.12. Rate constants for the reaction of **1-OTs** in DMSO in the presence of various amines

Nu	[1-OTs]$_o$/M	[Nu]/M	k_{obs}/s^{-1}	k_{obs} vs. [Nu] correlation	$k_2/M^{-1} s^{-1}$
DABCO	2.36 × 10^{-3}	0.00	6.76 × 10^{-6}	$y = 1.11E-04x + 6.01E-06$; $R^2 = 9.97E-01$	**1.11 × 10^{-4}**
	2.02 × 10^{-3}	0.25	3.29 × 10^{-5}		
	1.80 × 10^{-3}	0.47	5.64 × 10^{-5}		
	2.16 × 10^{-3}	0.62	7.72 × 10^{-5}		
piperidine	2.36 × 10^{-3}	0.00	6.76 × 10^{-6}	$y = 1.13E-04x + 7.92E-06$; $R^2 = 9.99E-01$	**1.13 × 10^{-4}**
	1.99 × 10^{-3}	0.20	3.20 × 10^{-5}		
	1.9662 × 10^{-3}	0.70	8.65 × 10^{-5}		
morpholine	2.36 × 10^{-3}	0.00	6.76 × 10^{-6}	$y = 5.80E-05x + 7.80E-06$; $R^2 = 9.95E-01$	**5.80 × 10^{-5}**
	2.27 × 10^{-3}	0.38	3.12 × 10^{-5}		
	4.17 × 10^{-3}	0.56	4.17 × 10^{-5}		
	5.49 × 10^{-3}	0.84	5.49 × 10^{-5}		
n-PrNH$_2$	2.36 × 10^{-3}	0.00	6.76 × 10^{-6}	$y = 5.77E-05x + 7.47E-06$; $R^2 = 9.96E-01$	**5.77 × 10^{-5}**
	1.79 × 10^{-3}	0.29	2.46 × 10^{-5}		
	1.83 × 10^{-3}	0.71	4.97 × 10^{-5}		
	1.86 × 10^{-3}	0.85	5.49 × 10^{-5}		

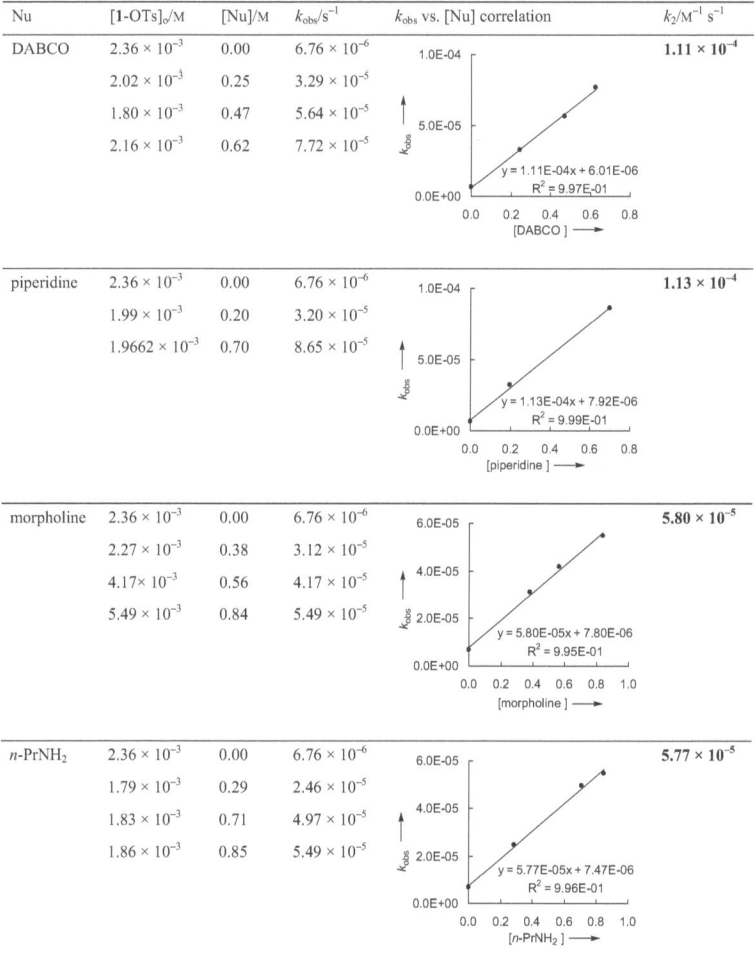

4. Leaving Group Dependence of the S_N1/S_N2 Ratio

4.5. References

(1) Hansch, C.; Leo, A.; Taft, R. W. *Chem. Rev.* **1991**, *91*, 165-195.
(2) Streidl, N.; Denegri, B.; Kronja, O.; Mayr, H. *Acc. Chem. Res.* **2010**, *43*, 1537-1549.
(3) Streidl, N.; Mayr, H. *Eur. J. Org. Chem.* **2009**, 2498-2506.
(4) Richard, J. P.; Toteva, M. M.; Amyes, T. L. *Org. Lett.* **2001**, *3*, 2225-2228.
(5) Lim, C.; Kim, S.-H.; Yoh, S.-D.; Fujio, M.; Tsuno, Y. *Tetrahedron Lett.* **1997**, *38*, 3243-3246.
(6) Kim, S. H.; Yoh, S.-D.; Fujio, M.; Imahori, H.; Mishima, M.; Tsuno, Y. *Bull. Korean Chem. Soc.* **1995**, *16*, 760-764.
(7) Liu, K.-T.; Chin, C.-P.; Lin, Y.-S.; Tsao, M.-L. *J. Chem. Res., Synop.* **1997**, 18-19.
(8) Christian, R. *Solvents and Solvent Effects in Organic Chemistry*; Wiley: Weinheim, 2003.
(9) Dvorko, G. F.; Ponomareva, E. A.; Ponomarev, M. E. *J. Phys. Org. Chem.* **2004**, *17*, 825-836.
(10) Dvorko, G. F.; Ponomareva, E. A.; Kulik, N. I. *Russ. Chem. Rev.* **1984**, *53*, 547-560.
(11) Hoz, S.; Basch, H.; Wolk, J. L.; Hoz, T.; Rozental, E. *J. Am. Chem. Soc.* **1999**, *121*, 7724-7725.
(12) Phan, T. B.; Breugst, M.; Mayr, H. *Angew. Chem.* **2006**, *118*, 3954-3959; *Angew. Chem. Int. Ed.* **2006**, *45*, 3869-3874.
(13) Mayr, H.; Patz, M. *Angew. Chem.* **1994**, *106*, 990-1010; Mayr, H.; Patz, M. *Angew. Chem. Int. Ed.* **1994**, *33*, 938-957.

5. Nucleofugality of Bromide in Other Aprotic Solvents

5.1. Introduction

In the absence of nucleophilic trapping agents potential ionization reactions of alkyl halides may occur reversibly and no gross reaction is observable. If the intermediately formed carbocation is trapped by an appropriate nucleophile, the rate of the ionization step can be measured. Typically, in protic solvents the solvent itself will react as nucleophile with the carbocation.

Scheme 5.1. S_N1 and S_N2 reaction pathway for the reaction of an alkyl halide.

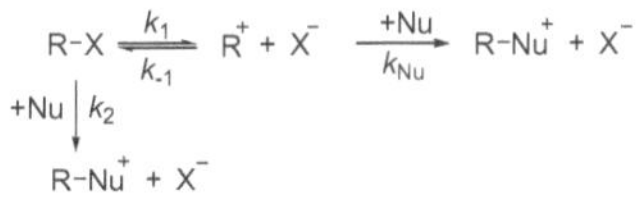

$$R\text{-}X \underset{k_{-1}}{\overset{k_1}{\rightleftharpoons}} R^+ + X^- \xrightarrow{\underset{k_{Nu}}{+Nu}} R\text{-}Nu^+ + X^-$$

$$+Nu \downarrow k_2$$

$$R\text{-}Nu^+ + X^-$$

If trapping is slow, i.e., if $k_{-1}[X^-] > k_{Nu}[Nu]$, common ion return will occur and the overall reaction rate constant, which is typically determined by conductometric or titrimetric measurements, does not reflect the ionization rate constant k_1. As the concentration of X^- increases during the course of the reaction, common ion return causes a deviation from the monoexponential rate law (eq. 5.1, for conductometric measurements) and the derivation of k_1 from the kinetic data becomes difficult.

$$G = G_\infty(1 - e^{-k_1 t}) + C \qquad (5.1)$$

To overcome these problems, a method to suppress common ion return by amines in protic solvents was previously developed.[1] This method was extended to the determination of heterolysis rate constants of benzhydryl chlorides in aprotic solvents.[2] So far this method was used to determine the nucleofugality parameters (N_f and s_f in eq. 5.2) of chloride in aprotic solvents and to derive electrofugality parameters of tritylium ions.[3]

$$\lg k_1(25 \,°\text{C}) = s_f(N_f + E_f) \qquad (5.2)$$

In the following, the application of this method on the reactions of benzhydryl bromides in various aprotic solvents will be reported.

5.2. Results and Discussion

A series of donor-substituted benzhydryl bromides (**9-15**)-Br were employed in this study. As presented in chapter 2, these benzhydryl bromides can easily be prepared from the corresponding benzhydrols by refluxing in neat acetyl bromide.

Scheme 5.2. Benzhydryl bromides employed in this study.

N°	X	Y
15	OMe	OMe
14	OMe	OPh
13	OMe	Me
12	OMe	H
11	Me	Me
10	Me	H
9	H	H

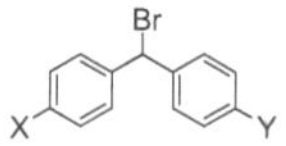

When benzhydryl bromides (**9-15**)-Br (Scheme 5.2) were dissolved in aprotic solvents in the presence of at least 10 eq. piperidine or N-methylpyrrolidine, a monoexponential increase of conductance according to equation 5.1 was observed, which allowed us to determine the first-order rate constants. Plots of the observed rate constants (k_{obs}) against the amine concentrations showed a linear increase of k_{obs} with [amine] at low concentrations of amine. At higher amine concentrations, the correlation lines bend downward and a maximum rate constant (k_{max}) is reached (Figure 5.1). When the amine concentration is further increased, the rate constant (k_{obs}) decreases. This decrease can be explained by a decrease of solvent polarity at high concentrations of amine.

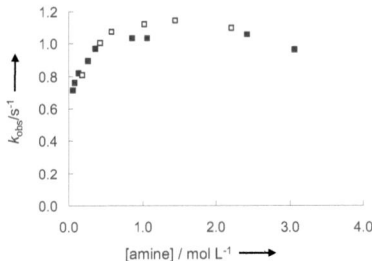

Figure 5.1. Plot of k_{obs} for the reaction of **12-Br** in acetonitrile in the presence of piperidine (■) and N-methylpyrrolidine (□) vs. [amine].

According to previous reports,[1-4] the maximum rate constants measured in aqueous or alcoholic solvents (Table 5.1) reflect the ionization rate constants (k_1) defined in scheme 5.1. In the following we will report on the reactions of benzhydryl bromides with amines in aprotic solvents.

Table 5.1. Observed maximum rate constants k_{max}/s^{-1} in various aprotic solvents using piperidine and N-methylpyrrolidine as trapping agents.

	MeCN		DMF		acetone	
	piperidine	N-methyl-pyrrolidine	piperidine	N-methyl-pyrrolidine	piperidine	N-methyl-pyrrolidine
15-Br					3.00×10^1	1.42×10^1
14-Br	6.27×10^1	5.22×10^1	1.76×10^1	1.30×10^1	2.44	1.24
13-Br	1.22×10^1	1.07×10^1	3.70	2.62	4.17×10^{-1}	2.00×10^{-1}
12-Br	1.03	1.14	4.62×10^{-1}	3.55×10^{-1}		
11-Br	1.08×10^{-1}	2.13×10^{-2}	6.09×10^{-2}	1.67×10^{-2}	1.09×10^{-2}	1.76×10^{-3}
10-Br	3.21×10^{-2}	6.96×10^{-3}	2.35×10^{-2}	2.37×10^{-3}	4.30×10^{-3}	3.13×10^{-4}
9-Br	1.21×10^{-3}	1.18×10^{-3}	1.04×10^{-2}	4.37×10^{-4}	2.07×10^{-3}	

In Figures 5.2-5.4 the maximum rate constants lg k_{max} for the reactions of **(9-15)-Br** in the presence of a high excess of piperidine or N-methylpyrrolidine were plotted against the electrofugalities E_f of the corresponding benzhydrylium ions.

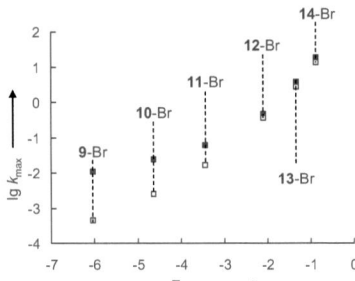

Figure 5.2. Plot of lg k_{max} at 25 °C for the ionization of benzhydryl bromides in the presence of piperidine (■) and N-methylpyrrolidine (□) in acetonitrile vs. the electrofugality parameter E_f.

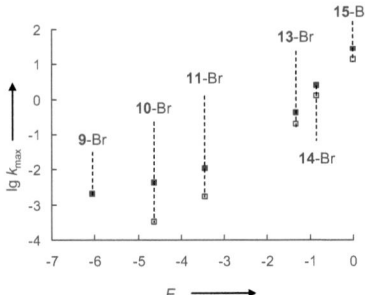

Figure 5.3. Plot of lg k_{max} at 25 °C for the ionization of benzhydryl bromides in the presence of piperidine (■) and N-methylpyrrolidine (□) in DMF vs. the electrofugality parameter E_f.

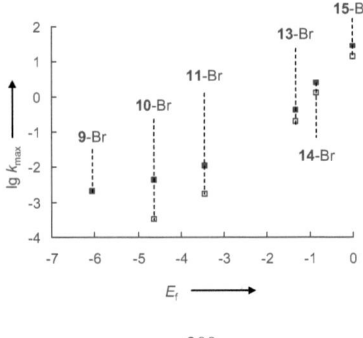

Figure 5.4. Plot of lg k_{max} at 25 °C for the ionization of benzhydryl bromides in the presence of piperidine (■) and N-methylpyrrolidine (□) in acetone vs. the electrofugality parameter E_f.

As expected from the analogous experiments with benzhydryl chlorides, the maximum rate constants k_{max} were nearly identical for the reactions of (**12-15**)-Br ($E_f \geq -2.09$) with piperidine and N-methylpyrrolidine. However, the maximum rate constants of the less reactive (**9-11**) Br ($E_f < -2.09$) with piperidine were significantly larger than the maximum rate constants obtained with N-methylpyrrolidine. Obviously, the maximum rate constants k_{max} become only independent of the type of amine additive, when the stabilized benzhydrylium ions **12$^+$-15$^+$** ($E_f \geq -2.09$) are formed. Presumably, the deactivated benzhydryl bromides (**9-11**)-Br ($E_f < -2.09$) react via a blend of S_N1 and S_N2 pathways with the amine. This interpretation is supported by the variable concentrations of piperidine that are needed to reach the maximum rate constants. As shown in Table 5.2, the less donor substituted benzhydryl bromides (**9-11**)-Br require significantly higher concentrations of piperidine to reach the maximum rate constant.

Table 5.2. Concentration of piperidine/M needed to reach k_{max} for the ionization reaction of benzhydryl bromides in a series of aprotic solvents.

N°-Br	[piperidine] in MeCN/M	[piperidine] in DMF/M	[piperidine] in acetone/M
9-Br	1.63	4.26	5.89
10-Br	1.61	2.89	4.31
11-Br	1.63-2.14[a]	3.95	3.71
12-Br	1.07	2.23	-
13-Br	0.83	1.44	1.51
14-Br	0.90	1.44	2.22
15-Br	-	-	2.60

[a] observed rate constants differ in this concentration range only by 1 %, see Table 5.7

Therefore, reactions which gave significantly different maximum rate constants with different amines, i.e., reactions with $k_{max} \leq 0.2$ s^{-1} were not used for the determination of the nucleofugality parameters N_f and s_f of bromide in a series of aprotic solvents (Figure 5.5).

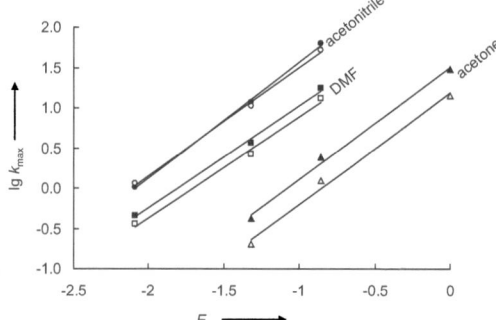

Figure 5.5. Plots of lg k_{max} at 25 °C for the ionization of benzhydryl bromides in the presence of piperidine (●,▲,■) and N-methylpyrrolidine (○,△,□) in various solvents vs. the electrofugality parameter E_f.

On the other hand, reactions with a ($k_{max} \geq 0.2$ s^{-1}) can be expected to react via S$_N$1 mechanism, and plots of k_{max} against the electrofugality parameter E_f resulted in linear correlations in agreement with equation 5.2. From these correlations, one can extract the nucleofugality parameters N_f as the negative intercepts on the abscissa (E_f axis) and the s_f parameters as the slopes of the correlation lines (Table 5.3). In previous investigations in protic[1,3] and aprotic solvents[2] piperidine was used as the standard nucleophile to react with the benzhydrylium ions formed by ionization of benzhydryl chlorides. Other amines were also used in some experiments to demonstrate that the maximum rate constant is almost independent of the type of amine. The maximum rate constant (k_{max}) for the ionization of chloro-bis(4-methoxyphenyl)methane (**15**-Cl) in neat acetonitrile obtained with piperidine differed by a factor of 1.6 from the maximum rate constant obtained with pyridine.[2] In 90 % aqueous acetone (90A10W) maximum rate constants for chloro-bis-(4-methoxyphenyl)methane (**15**-Cl) obtained with piperidine differed by factors of 1.4 to 1.8 from the rate constants obtained with other nucleophilic amines.[1] In case of the benzhydryl bromides (**12-15**)-Br ($E_f \geq -2.09$) investigated in this work the maximum rate constants k_{max} obtained with piperidine and N-methylpyrrolidine differed by factors of 0.9 to 2.1. Therefore, it is suggested that the maximum rate constants presented in Table 5.1 correspond to the rates of ionization of benzhydryl bromides ($E_f \geq -2.09$) and can be used to determine nucleofugality parameters (Table 5.3) of bromide in these solvents.

Table 5.3. Nucleofugality parameters of bromide in aprotic solvents determined by the "amine method" using piperidine and N-methylpyrrolidine.[a]

N_f / s_f	acetonitrile	DMF	acetone
piperidine	2.09 / 1.45	1.81 / 1.27	1.08 / 1.39
N-methylpyrrolidine	2.12 / 1.34	1.71 / 1.26	0.86 / 1.38

[a] Ideally N_f and s_f should not depend on the nature of the trapping amine.

For the less donor substituted benzhydryl bromides ($E_f < -2.09$) the maximum rate constants observed with piperidine are approximately one order of magnitude larger than the maximum rate constants determined with N-methylpyrrolidine. Presumably, the ionization of the benzhydryl bromide followed by rapid trapping of the benzhydrylium ion by the amine is not the predominant reaction mechanism; instead piperidine reacts via an S_N2 reaction. Now, the question arises why reactivity maxima are reached at certain concentrations of amines, i.e., why an increase of the concentration of piperidine does not result in an acceleration of the reaction when the concentration of piperidine exceeds 1 mol L^{-1}. In the preceding chapters it was demonstrated that in DMSO the two reaction mechanisms (S_N1 and S_N2) can be clearly separated. Thus, the reaction of the monomethyl substituted benzhydryl bromide **10**-Br in DMSO was employed to investigate the second-order rate constants k_2 at high amine concentrations of piperidine. As depicted in Figure 5.6 we have studied these reactions also at higher concentrations of piperidine and found that the observed rate constants k_{obs} for the reaction of **10**-Br with piperidine in DMSO at high amine concentrations ([piperidine] ≥ 1.09 M) deviate from the linear correlation observed at low amine concentrations.

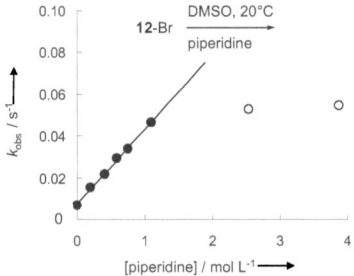

Figure 5.6. Plot of k_{obs} for the reaction of **10**-Br in DMSO in the presence of piperidine and vs. [piperidine].

As a concentration of [piperidine] = 2 M corresponds to 19.8 vol.% piperidine in DMSO, one can assume that in this concentration range variation of [amine] causes a significant change of solvent polarity and thus inhibits a further acceleration of the S_N2 reaction.

Assuming that variations of [amine] at concentrations below 1 M (10 vol.% piperidine) have neglible effects on solvent polarity (as demonstrated by the linear correlations in Figure 3.2 on p. 123), second-order rate constants for the S_N2 reactions in different solvents can be derived from plots of k_{obs} vs. [piperidine] in this concentration range (Figure 5.7, Table 5.4).

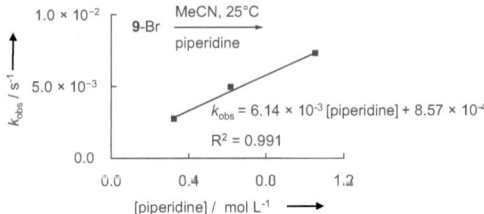

Figure 5.7. Plot of k_{obs} for the reaction of **9-Br** in acetonitrile in the presence of piperidine vs. the concentration of piperidine (for other plots see Experimental Section).

Table 5.4. Observed second order rate constants for the reactions of benzhydryl bromides with piperidine in different solvents.

$k_2/\text{s}^{-1}\,\text{M}^{-1}$	9-Br	10-Br	11-Br
DMSO^a	1.69×10^{-2}	3.57×10^{-2}	-
MeCN	6.14×10^{-3}	1.75×10^{-2}	7.34×10^{-2}
DMF	5.00×10^{-3}	9.79×10^{-3}	2.63×10^{-2}
acetone	8.22×10^{-4}	1.69×10^{-3}	4.31×10^{-3}

^{a} DMSO data is taken from Chapter 3.

When these second-order rate constants were plotted against Hammett's σ-values, reaction constants $\rho = -3.17$ to -2.12 (-1.91 for DMSO) were obtained (Figure 5.8). In contrast to the results in Chapter 3 (Figure 3.3), linear correlations were obtained. This is probably an effect of using only donor substituents (methyl).

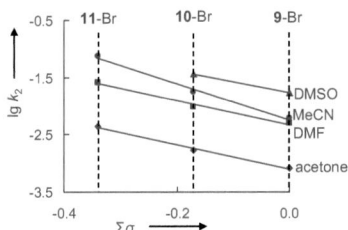

Figure 5.8. Plot of lg k_2 for the reactions of the benzhydryl bromides with piperidine in DMSO ; lg $k_2 = -1.91\sigma - 1.77$ (▲), acetonitrile; lg $k_2 = -3.17\sigma - 2.24$ (●), DMF; lg $k_2 = -2.12\sigma - 2.32$ (■) and acetone; lg $k_2 = -2.12\sigma - 3.10$ (♦) vs. Hammett's substituent constants σ.[5]

At a first glance these reaction constants ρ seem to be considerably more negative than those for the S_N2 reactions investigated in Chapter 4 (−0.81 to −0.21). However, when looking at the two data points for the S_N2 reaction of **10**-Br and **9**-Br with piperidine in DMSO a reactivity ratio of 12, corresponding to a slope of $\rho = -1.91$ (Figure 5.8 and 4.4) is observed. Thus, donor substituents appear to accelerate the S_N2 reactions significantly while acceptor substituents have only a weak decelerating effect. Tsuno *et. al* observed a similar behavior for the reaction of substituted 1-arylethyl bromides with pyridine.[6] In their work, donor substituents strongly accelerate the S_N2 reaction, while acceptor substituents show only a weak influence on the reaction rate. This resulted in concave Hammett plots. Tsuno explained this behavior by a change from a loose cationic transition state for the reaction of donor-substituted, to a tight transition state for the reaction of acceptor-substituted 1-arylethyl bromides. Therefore, the S_N2 reactivity for acceptor substituted systems is higher than one could expect from the extrapolation of Hamett correlation.

5.3. Conclusion and Outlook

The straightforward determination of ionization rates of benzhydryl chlorides by trapping the intermediate carbocations by amines cannot easily be employed for measuring the ionization rates of benzhydryl bromides in solvents of low nucleophilicity (acetonitrile, DMF, acetone). As bromide is a significantly better nucleofuge than chloride in these solvents, conveniently measurable ionization rates can only be achieved with less donor substituted benzhydrylium bromides (Figure 5.9), i.e., systems which tend to react via S_N2 process (Richard Jencks life-time argument).

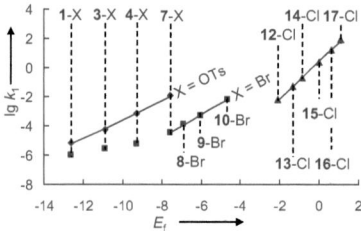

Figure 5.9. Plot of lg k_1 for the solvolysis of various substituted benzhydryl chlorides (at 25 °C), bromides and tosylates in DMSO at 20 °C vs. electrofugality E_f.

206

Nevertheless, for some benzhydryl bromides, ionization rates (first-order rate constants) could be determined, and Figure 5.10 shows that in dipolar aprotic solvents bromide is a 400-800 times better nucleofuge than chloride.

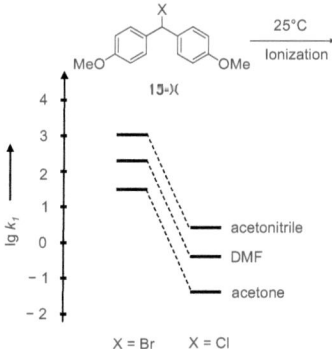

Figure 5.10. Comparison of calculated lg k_1 for the ionizations of the dimethoxy substituted benzhydryl chloride (15-Cl) and bromide (15-Br) in three different aprotic solvents. All rate constants were determined by the "amine method" using piperidine.

This ratio is considerably higher than in many aqueous and alcoholic solvents where alkyl bromides typically ionize about one order of magnitude faster than alkyl chlorides. The high k_{Br}/k_{Cl} ratio in dipolar aprotic solvents combined with the observation that both classes of compounds ionize with almost equal rates in trifluoroethanol illustrate that hydrogen bonding plays a much greater role for the ionization of chlorides than of bromides.

5.4. Experimental Section

When benzhydryl bromides (9-15)-Br were dissolved in aprotic media in the presence of piperidine or N-methylpyrrolidine, an increase of conductance was observed. A calibration experiment, i.e., stepwise addition of the rapidly ionizing benzhydryl bromide 12-Br to acetonitrile containing 0.5 M N-methylpyrrolidine, showed a linear correlation between the initial concentration of the benzhydryl bromide 12-Br and the conductance at the end of the reaction within the investigated concentration range. Consequently, monoexponential increases of the conductance (G) were observed during the solvolysis reaction and the first-order rate constants k_1 (Table 2) were obtained by fitting the time dependent conductance G to the monoexponential function (eq. 5.1).

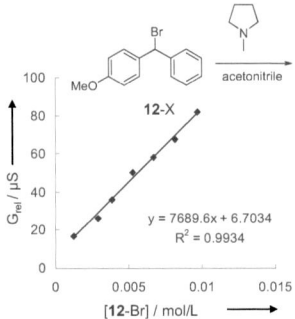

Figure 5.11. Conductance at t_∞ vs. concentration of **12-Br** in acetonitrile at 25 °C. After the addition of a portion of **12-Br**, the next conductance value was taken when the conductance remained constant.

Maximum rate constants k_{max} were obtained by plotting the observed rate constants k_{obs} against the amine concentrations (Tables 5.4-5.21). Maximum rate constants were typically observed at an amine concentration of 1.0 M. The temperature of the solutions during all kinetic studies was kept constant at 25.0 °C (± 0.1 °C) by using a circulating bath thermostat. Slow reactions ($k_{obs} < 1.1 \times 10^{-1}$) were monitored with a conventional conductometer (conductometers: Radiometer Analytical CDM 230 or Tacussel CD 810, Pt electrode: WTW LTA 1/NS). The final concentrations of benzhydryl bromides were around 2.5 - 3.0 × 10^{-3} M. Fast ionization reactions, were measured with a stopped-flow conductometer (Hi-Tech Scientific SF-61 DX2, platinum electrodes, cell volume: 21 µL, cell constant 4.24 cm^{-1}, minimum dead time 2.2 ms) Final concentrations of benzhydryl bromide were 5.0 - 7.0 × 10^{-3} M for the stopped flow measurements. After injection of the benzhydryl derivative into the ionizing medium, an increase of conductance was observed, that was recorded at certain time intervals resulting in about 3000 data points for each measurement. The first-order rate constants k_1 (s^{-1}) were obtained by least squares fitting of the conductance data to the single-exponential equation (5.3).

Table 5.5. Individual observed rate constants at 25 °C for the reaction of **9-Br** in acetonitrile in the presence of piperidine (■) and N-methylpyrrolidine (□).

9-Br in acetonitrile		
[piperidine]/M	k_{obs}/s^{-1}	k_{obs} vs. [amine] correlation
0.33	2.73×10^{-3}	
0.62	4.91×10^{-3}	
1.06	7.25×10^{-3}	
1.60	9.30×10^{-3}	
1.63	9.50×10^{-3}	$y = 6.14\text{E-}03x + 8.57\text{E-}04$
2.16	1.14×10^{-2}	$R^2 = 9.91\text{E-}01$
2.44	1.20×10^{-2}	
3.74	1.21×10^{-3}	
5.20	1.16×10^{-3}	
[N-methylpyrrolidine]/M	k_{obs}/s^{-1}	
0.78	8.38×10^{-4}	
1.02	8.96×10^{-4}	
1.71	1.12×10^{-3}	
1.94	1.18×10^{-3}	
2.77	1.14×10^{-3}	

Table 5.6. Individual observed rate constants at 25 °C for the reaction of **10-Br** in acetonitrile in the presence of piperidine (■) and N-methylpyrrolidine (□).

10-Br in acetonitrile		
[piperidine]/M	k_{obs}/s^{-1}	k_{obs} vs. [amine] correlation
0.26	6.79×10^{-3}	
0.53	1.24×10^{-2}	
0.83	1.79×10^{-2}	
1.10	2.16×10^{-2}	
1.20	2.36×10^{-2}	
1.61	2.89×10^{-2}	
2.86	3.21×10^{-2}	$y = 1.77\text{E-}02x + 2.61\text{E-}03$
4.37	3.04×10^{-2}	$R^2 = 9.92\text{E-}01$
[N-methylpyrrolidine]/M	k_{obs}/s^{-1}	
0.56	3.98×10^{-3}	
0.89	4.80×10^{-3}	
1.45	6.16×10^{-3}	
2.06	6.96×10^{-3}	
2.63	6.25×10^{-3}	

Table 5.7. Individual observed rate constants at 25 °C for the reaction of **11-Br** in acetonitrile in the presence of piperidine (■) and N-methylpyrrolidine (□).

11-Br in acetonitrile		
[piperidine]/M	k_{obs}/s^{-1}	k_{obs} vs. [amine] correlation
0.13	2.26×10^{-2}	
0.24	3.08×10^{-2}	
0.44	4.70×10^{-2}	
0.77	7.25×10^{-2}	
1.16	9.72×10^{-2}	
1.63	1.07×10^{-1}	$y = 7.34\text{E-}02x + 1.40\text{E-}02$
2.14	1.08×10^{-1}	$R^2 = 9.97\text{E-}01$
[N-methylpyrrolidine]/M	k_{obs}/s^{-1}	
0.30	1.32×10^{-2}	
0.70	1.70×10^{-2}	
1.20	1.97×10^{-2}	
1.60	2.13×10^{-2}	
2.00	2.13×10^{-2}	

Table 5.8. Individual observed rate constants at 25 °C for the reaction of **12-Br** in acetonitrile in the presence of piperidine (■) and N-methylpyrrolidine (□).

12-Br in acetonitrile		
[piperidine]/M	k_{obs}/s^{-1}	k_{obs} vs. [amine] correlation
0.07	7.10×10^{-1}	
0.09	7.55×10^{-1}	
0.13	8.14×10^{-1}	
0.26	8.94×10^{-1}	
0.37	9.64×10^{-1}	
0.87	1.03	
1.07	1.03	
2.44	1.02	
3.07	9.60×10^{-1}	
[N-methylpyrrolidine]/M	k_{obs}/s^{-1}	
0.19	8.06×10^{-1}	
0.43	1.00	
0.59	1.07	
1.03	1.12	
1.45	1.14	
2.22	1.10	

Table 5.9. Individual observed rate constants at 25 °C for the reaction of **13-Br** in acetonitrile in the presence of piperidine (■) and N-methylpyrrolidine (□).

13-Br in acetonitrile		
[piperidine]/M	k_{obs}/s^{-1}	k_{obs} vs. [amine] correlation
0.05	7.66	
0.13	9.03	
0.26	9.68	
0.83	1.22×10^1	
1.53	1.22×10^1	
1.95	1.18×10^1	
[N-methylpyrrolidine]/M	k_{obs}/s^{-1}	
0.19	7.91	
0.43	9.77	
0.59	1.03×10^1	
1.03	1.06×10^1	
1.45	1.07×10^1	
2.22	9.56	

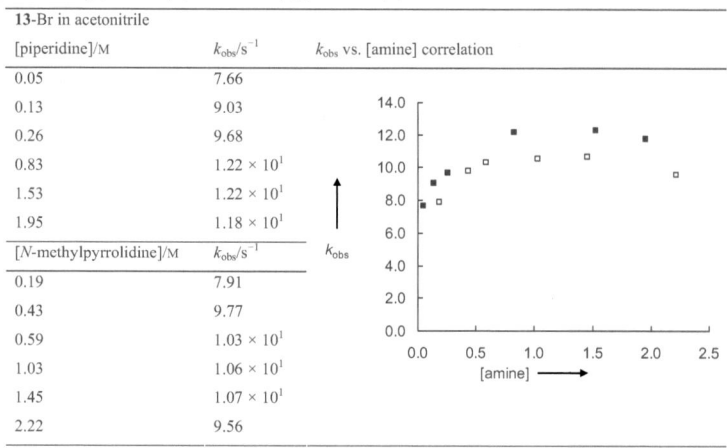

Table 5.10. Individual observed rate constants at 25 °C for the reaction of **14-Br** in acetonitrile in the presence of piperidine (■) and N-methylpyrrolidine (□).

14-Br in acetonitrile		
[piperidine]/M	k_{obs}/s^{-1}	k_{obs} vs. [amine] correlation
0.08	4.36×10^1	
0.23	5.15×10^1	
0.57	5.89×10^1	
0.90	6.27×10^1	
1.24	6.15×10^1	
1.67	5.98×10^1	
2.12	5.65×10^1	
[N-methylpyrrolidine]/M	k_{obs}/s^{-1}	
0.19	4.19×10^1	
0.43	4.95×10^1	
0.59	5.12×10^1	
1.03	5.12×10^1	
1.45	5.22×10^1	
2.22	5.19×10^1	

Table 5.11. Individual observed rate constants at 25 °C for the reaction of **9-Br** in DMF in the presence of piperidine (■) and N-methylpyrrolidine (□).

9-Br in DMF		
[piperidine]/M	k_{obs}/s^{-1}	k_{obs} vs. [amine] correlation
0.36	2.31×10^{-3}	
0.60	3.68×10^{-3}	
0.93	5.20×10^{-3}	
1.16	6.37×10^{-3}	
1.52	7.40×10^{-3}	
2.50	9.66×10^{-3}	
4.26	1.04×10^{-2}	y = 5.00E-03x + 5.88E-04
4.92	9.96×10^{-3}	R^2 = 9.99E-01
[N-methylpyrrolidine]/M	k_{obs}/s^{-1}	
0.42	3.39×10^{-4}	
0.92	3.39×10^{-4}	
1.48	4.16×10^{-4}	
2.30	4.37×10^{-4}	
3.68	3.47×10^{-4}	

Table 5.12. Individual observed rate constants at 25 °C for the reaction of **10-Br** in DMF in the presence of piperidine (■) and N-methylpyrrolidine (□).

10-Br in DMF		
[piperidine]/M	k_{obs}/s^{-1}	k_{obs} vs. [amine] correlation
0.31	5.12×10^{-3}	
0.61	8.96×10^{-3}	
0.94	1.19×10^{-2}	
1.27	1.46×10^{-2}	
1.55	1.65×10^{-2}	
1.82	1.81×10^{-2}	
2.15	2.05×10^{-2}	
2.89	2.35×10^{-2}	$y = 9.79\text{E-}03x + 2.47\text{E-}03$
3.26	2.33×10^{-2}	$R^2 = 9.90\text{E-}01$
4.89	1.53×10^{-2}	
[N-methylpyrrolidine]/M	k_{obs}/s^{-1}	
0.54	1.86×10^{-3}	
0.85	2.03×10^{-3}	
1.31	2.30×10^{-3}	
1.97	2.37×10^{-3}	
2.27	2.27×10^{-3}	

Table 5.13. Individual observed rate constants at 25 °C for the reaction of **11**-Br in DMF in the presence of piperidine (■) and N-methylpyrrolidine (□).

11-Br in DMF		
[piperidine]/M	k_{obs}/s^{-1}	k_{obs} vs. [amine] correlation
0.36	2.33×10^{-2}	
0.56	2.90×10^{-2}	
0.81	3.68×10^{-2}	
1.18	4.49×10^{-2}	
1.97	5.36×10^{-2}	
2.66	5.81×10^{-2}	
3.95	6.09×10^{-2}	
5.31	5.26×10^{-2}	
[N-methylpyrrolidine]/M	k_{obs}/s^{-1}	
0.50	1.45×10^{-2}	
0.86	1.58×10^{-2}	
1.16	1.66×10^{-2}	
1.62	1.67×10^{-2}	
2.32	1.62×10^{-2}	
4.03	1.28×10^{-2}	

Plot: $y = 2.63\text{E-}02x + 1.43\text{E-}02$, $R^2 = 9.94\text{E-}01$

Table 5.14. Individual observed rate constants at 25 °C for the reaction of **12**-Br in DMF in the presence of piperidine (■) and N-methylpyrrolidine (□).

12-Br in DMF		
[piperidine]/M	k_{obs}/s^{-1}	k_{obs} vs. [amine] correlation
0.45	3.29×10^{-1}	
0.60	3.69×10^{-1}	
0.98	4.24×10^{-1}	
1.44	4.56×10^{-1}	
2.23	4.62×10^{-1}	
2.54	4.38×10^{-1}	
[N-methylpyrrolidine]/M	k_{obs}/s^{-1}	
0.16	2.84×10^{-1}	
0.45	3.33×10^{-1}	
0.74	3.55×10^{-1}	
1.46	3.46×10^{-1}	
1.84	3.20×10^{-1}	

Table 5.15. Individual observed rate constants at 25 °C for the reaction of **13-Br** in DMF in the presence of piperidine (■) and N-methylpyrrolidine (□).

13-Br in DMF		
[piperidine]/M	k_{obs}/s^{-1}	k_{obs} vs. [amine] correlation
0.45	3.01	
0.60	3.18	
0.98	3.48	
1.44	3.70	
2.23	3.62	
2.54	3.46	
[N-methylpyrrolidine]/M	k_{obs}/s^{-1}	
0.16	2.10	
0.45	2.51	
0.74	2.62	
1.46	2.49	
1.84	2.2	

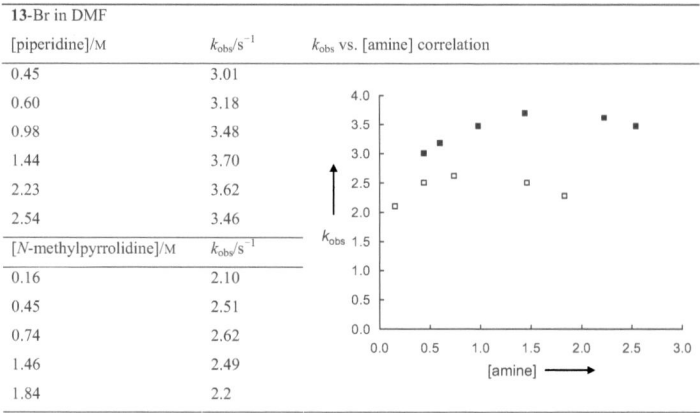

Table 5.16. Individual observed rate constants at 25 °C for the reaction of **14-Br** in DMF in the presence of piperidine (■) and N-methylpyrrolidine (□).

14-Br in DMF		
[piperidine]/M	k_{obs}/s^{-1}	k_{obs} vs. [amine] correlation
0.18	1.19×10^1	
0.45	1.44×10^1	
0.60	1.53×10^1	
0.98	1.67×10^1	
1.44	1.76×10^1	
2.23	1.74×10^1	
2.54	1.66×10^1	
[N-methylpyrrolidine]/M	k_{obs}/s^{-1}	
0.16	1.06×10^1	
0.45	1.27×10^1	
0.74	1.30×10^1	
1.46	1.23×10^1	
1.84	1.15×10^1	

Table 5.17. Individual observed rate constants at 25 °C for the reaction of **9-Br** in acetone in the presence of piperidine (■) and N-methylpyrrolidine (□).

9-Br in acetone		
[piperidine]/M	k_{obs}/s^{-1}	k_{obs} vs. [amine] correlation
0.34	2.26×10^{-4}	
0.63	4.37×10^{-4}	
0.75	5.74×10^{-4}	
1.15	8.87×10^{-4}	
2.53	1.47×10^{-4}	
3.60	1.87×10^{-4}	
4.47	2.01×10^{-4}	
5.89	2.07×10^{-4}	

(correlation plot: $y = 8.22\text{E-}04x - 5.92\text{E-}05$, $R^2 = 9.96\text{E-}01$)

Table 5.18. Individual observed rate constants at 25 °C for the reaction of **10-Br** in acetone in the presence of piperidine (■) and N-methylpyrrolidine (□).

10-Br in acetone		
[piperidine]/M	k_{obs}/s^{-1}	k_{obs} vs. [amine] correlation
0.25	4.17×10^{-4}	
0.53	8.84×10^{-4}	
0.77	1.34×10^{-3}	
1.24	2.07×10^{-3}	
1.55	2.64×10^{-3}	
1.88	3.19×10^{-3}	
2.81	4.00×10^{-3}	
3.34	4.28×10^{-3}	
4.31	4.30×10^{-3}	
[N-methylpyrrolidine]/M	k_{obs}/s^{-1}	
0.59	1.93×10^{-4}	
0.80	2.31×10^{-4}	
1.44	2.90×10^{-4}	
1.82	3.13×10^{-4}	
2.54	2.98×10^{-4}	

(correlation plot: $y = 1.69\text{E-}03x + 3.74\text{E-}06$, $R^2 = 9.99\text{E-}01$)

Table 5.19. Individual observed rate constants at 25 °C for the reaction of **11-Br** in acetone in the presence of piperidine (■) and N-methylpyrrolidine (□).

11-Br in acetone

[piperidine]/M	k_{obs}/s^{-1}	k_{obs} vs. [amine] correlation
0.34	1.66×10^{-3}	
0.60	2.83×10^{-3}	
0.86	4.01×10^{-3}	
1.24	5.55×10^{-3}	
1.69	7.66×10^{-3}	
1.99	8.33×10^{-3}	
2.42	9.61×10^{-3}	
3.71	1.09×10^{-2}	
4.77	1.08×10^{-2}	
[N-methylpyrrolidine]/M	k_{obs}/s^{-1}	
0.86	1.32×10^{-3}	
1.07	1.32×10^{-3}	
1.43	1.74×10^{-3}	
1.95	1.76×10^{-3}	
3.01	1.64×10^{-3}	
3.94	1.32×10^{-3}	

Plot: k_{obs} vs [amine]; linear fit $y = 4.31\text{E-}03x + 2.35\text{E-}04$, $R^2 = 1.00\text{E}+00$.

Table 5.20. Individual observed rate constants at 25 °C for the reaction of **13**-Br in acetone in the presence of piperidine (■) and N-methylpyrrolidine (□).

13-Br in acetone		
[piperidine]/M	k_{obs}/s^{-1}	k_{obs} vs. [amine] correlation
0.08	6.86×10^{-2}	
0.18	1.15×10^{-1}	
0.57	2.44×10^{-1}	
1.40	3.99×10^{-1}	
1.45	4.23×10^{-1}	
1.51	4.17×10^{-1}	
2.60	3.56×10^{-1}	
3.52	2.98×10^{-1}	
[N-methylpyrrolidine]/M	k_{obs}/s^{-1}	
0.13	6.56×10^{-2}	
0.39	1.31×10^{-1}	
0.83	1.85×10^{-1}	
1.53	2.00×10^{-1}	
2.19	1.79×10^{-1}	

Table 5.21. Individual observed rate constants at 25 °C for the reaction of **14**-Br in acetone in the presence of piperidine (■) and N-methylpyrrolidine (□).

14-Br in acetone		
[piperidine]/M	k_{obs}/s^{-1}	k_{obs} vs. [amine] correlation
0.28	8.52×10^{-1}	
0.59	1.32	
1.03	1.77	
1.45	2.12	
2.22	2.44	
2.67	2.26	
[N-methylpyrrolidine]/M	k_{obs}/s^{-1}	
0.13	4.16×10^{-1}	
0.39	8.42×10^{-1}	
0.83	1.10	
1.53	1.24	
2.19	1.13	

Table 5.22. Individual observed rate constants at 25 °C for the reaction of **15**-Br in acetone in the presence of piperidine (■) and N-methylpyrrolidine (□).

15-Br in acetone		
[piperidine]/M	k_{obs}/s^{-1}	k_{obs} vs. [amine] correlation
0.08	6.40	
0.18	9.50	
0.34	1.38×10^1	
0.57	1.85×10^1	
0.86	2.26×10^1	
1.40	2.71×10^1	
1.51	2.81×10^1	
2.60	3.00×10^1	
3.52	2.88×10^1	
[N-methylpyrrolidine]/M	k_{obs}/s^{-1}	
0.13	5.55	
0.39	9.69	
0.83	1.30×10^1	
1.53	1.42×10^1	
2.19	1.31×10^1	

5.5. References

(1) Streidl, N.; Antipova, A.; Mayr, H. *J. Org. Chem.* **2009**, *74*, 7328-7334
(2) Streidl, N.; Mayr, H. *Eur. J. Org. Chem.* **2009**, 2498-2506
(3) Horn, M.; Mayr, H. *Eur. J. Org. Chem.* **2011**, Early view.
(4) Streidl, N., Denegri, D.; Kronja, O.; Mayr, H. *Acc. Chem. Res.* **2010**, *43*, 1537-1549.
(5) Hansch, C.; Leo, A.; Taft, R. W. *Chem. Rev.* **1991**, *91*, 165-195
(6) Lim, C.; Kim, S.-H.; Yoh, S.-D.; Fujio, M.; Tsuno, Y. *Tetrahedron Letters* **1997**, *38*, 3243-3246.
(7) Baidya, M.; Kobayashi, S.; Mayr, H. *J. Am. Chem. Soc.* **2010**, *132*, 4796-4805

i want morebooks!

Buy your books fast and straightforward online - at one of world's fastest growing online book stores! Environmentally sound due to Print-on-Demand technologies.

Buy your books online at

www.get-morebooks.com

Kaufen Sie Ihre Bücher schnell und unkompliziert online – auf einer der am schnellsten wachsenden Buchhandelsplattformen weltweit! Dank Print-On-Demand umwelt- und ressourcenschonend produziert.

Bücher schneller online kaufen

www.morebooks.de

VDM Verlagsservicegesellschaft mbH
Heinrich-Böcking-Str. 6-8   Telefon: +49 681 3720 174   info@vdm-vsg.de
D - 66121 Saarbrücken   Telefax: +49 681 3720 1749   www.vdm-vsg.de

MIX
Papier aus verantwortungsvollen Quellen
Paper from responsible sources
FSC® C105338
www.fsc.org

Printed by Books on Demand GmbH, Norderstedt / Germany